T0229668

Leading Effective Virtual Teams

Overcoming Time and Distance to Achieve Exceptional Results

OTHER TELECOMMUNICATIONS BOOKS FROM AUERBACH

Ad Hoc Mobile Wireless Networks: Principles, Protocols, and Applications
Subir Kumar Sarkar, T.G. Basavaraju, and C. Puttamadappa
ISBN 978-1-4200-6221-2

Communication and Networking in Smart Grids
Yang Xiao (Editor)
ISBN 978-1-4398-7873-6

Delay Tolerant Networks: Protocols and Applications
Athanasios V. Vasilakos, Yan Zhang, and Thrasyvoulos Spyropoulos
ISBN 978-1-4398-1108-5

Emerging Wireless Networks: Concepts, Techniques and Applications
Christian Makaya and Samuel Pierre (Editors)
ISBN 978-1-4398-2135-0

Game Theory in Communication Networks: Cooperative Resolution of Interactive Networking Scenarios
Josephina Antoniou and Andreas Pitsillides
ISBN 978-1-4398-4808-1

Green Communications: Theoretical Fundamentals, Algorithms and Applications
Jinsong Wu, Sundeep Rangan, and Honggang Zhang
ISBN 978-1-4665-0107-2

Green Communications and Networking
F. Richard Yu, Xi Zhang, and Victor C.M. Leung (Editors)
ISBN 978-1-4398-9913-7

Green Mobile Devices and Networks: Energy Optimization and Scavenging Techniques
Hrishikesh Venkataraman and Gabriel-Miro Muntean (Editors)
ISBN 978-1-4398-5989-6

Handbook on Mobile Ad Hoc and Pervasive Communications
Laurence T. Yang, Xingang Liu, and Mieso K. Denko (Editors)
ISBN 978-1-4398-4616-2

IP Telephony Interconnection Reference: Challenges, Models, and Engineering
Mohamed Boucadair, Isabel Borges, Pedro Miguel Neves, and Olafur Pall Einarsson
ISBN 978-1-4398-5178-4

LTE-Advanced Air Interface Technology
Xincheng Zhang and Xiaojin Zhou
ISBN 978-1-4665-0152-2

Media Networks: Architectures, Applications, and Standards
Hassnaa Moustafa and Sherali Zeadally (Editors)
ISBN 978-1-4398-7728-9

Multihomed Communication with SCTP (Stream Control Transmission Protocol)
Victor C.M. Leung, Eduardo Parente Ribeiro, Alan Wagner, and Janardhan Iyengar
ISBN 978-1-4665-6698-9

Multimedia Communications and Networking
Mario Marques da Silva
ISBN 978-1-4398-7484-4

Near Field Communications Handbook
Syed A. Ahson and Mohammad Ilyas (Editors)
ISBN 978-1-4200-8814-4

Next-Generation Batteries and Fuel Cells for Commercial, Military, and Space Applications
A. R. Jha, ISBN 978-1-4398-5066-4

Physical Principles of Wireless Communications, Second Edition
Victor L. Granatstein, ISBN 978-1-4398-7897-2

Security of Mobile Communications
Noureddine Boudriga, ISBN 978-0-8493-7941-3

Smart Grid Security: An End-to-End View of Security in the New Electrical Grid
Gilbert N. Sorebo and Michael C. Echols
ISBN 978-1-4398-5587-4

Transmission Techniques for 4G Systems
Mário Marques da Silva
ISBN 978-1-4665-1233-7

Transmission Techniques for Emergent Multicast and Broadcast Systems
Mário Marques da Silva, Americo Correia, Rui Dinis, Nuno Souto, and Joao Carlos Silva
ISBN 978-1-4398-1593-9

TV Content Analysis: Techniques and Applications
Yiannis Kompatsiaris, Bernard Merialdo, and Shiguo Lian (Editors)
ISBN 978-1-4398-5560-7

TV White Space Spectrum Technologies: Regulations, Standards, and Applications
Rashid Abdelhaleem Saeed and Stephen J. Shellhammer
ISBN 978-1-4398-4879-1

Wireless Sensor Networks: Current Status and Future Trends
Shafiullah Khan, Al-Sakib Khan Pathan, and Nabil Ali Alrajeh
ISBN 978-1-4665-0606-0

Wireless Sensor Networks: Principles and Practice
Fei Hu and Xiaojun Cao
ISBN 978-1-4200-9215-8

AUERBACH PUBLICATIONS
www.auerbach-publications.com
To Order Call: 1-800-272-7737 • Fax: 1-800-374-3401
E-mail: orders@crcpress.com

Leading Effective Virtual Teams

Overcoming Time and Distance to Achieve Exceptional Results

Nancy M. Settle-Murphy

CRC Press
Taylor & Francis Group
Boca Raton   London   New York

CRC Press is an imprint of the
Taylor & Francis Group, an **Informa** business

AN AUERBACH BOOK

guided**insights**

CRC Press
Taylor & Francis Group
6000 Broken Sound Parkway NW, Suite 300
Boca Raton, FL 33487-2742

© 2013 by Taylor & Francis Group, LLC
CRC Press is an imprint of Taylor & Francis Group, an Informa business

No claim to original U.S. Government works

Version Date: 20121026

International Standard Book Number: 978-1-4665-5786-4 (Paperback)

This book contains information obtained from authentic and highly regarded sources. Reasonable efforts have been made to publish reliable data and information, but the author and publisher cannot assume responsibility for the validity of all materials or the consequences of their use. The authors and publishers have attempted to trace the copyright holders of all material reproduced in this publication and apologize to copyright holders if permission to publish in this form has not been obtained. If any copyright material has not been acknowledged please write and let us know so we may rectify in any future reprint.

Except as permitted under U.S. Copyright Law, no part of this book may be reprinted, reproduced, transmitted, or utilized in any form by any electronic, mechanical, or other means, now known or hereafter invented, including photocopying, microfilming, and recording, or in any information storage or retrieval system, without written permission from the publishers.

For permission to photocopy or use material electronically from this work, please access www.copyright.com (http://www.copyright.com/) or contact the Copyright Clearance Center, Inc. (CCC), 222 Rosewood Drive, Danvers, MA 01923, 978-750-8400. CCC is a not-for-profit organization that provides licenses and registration for a variety of users. For organizations that have been granted a photocopy license by the CCC, a separate system of payment has been arranged.

Trademark Notice: Product or corporate names may be trademarks or registered trademarks, and are used only for identification and explanation without intent to infringe.

Visit the Taylor & Francis Web site at
http://www.taylorandfrancis.com

and the CRC Press Web site at
http://www.crcpress.com

I dedicate this book to my friend, colleague, collaborator, and teacher, Julia Young. Thanks to our many collaborations over the last several years—despite our distance of 3,000 miles—I've mastered the art and science of planning and designing engaging virtual conversations. Julia is not only one of the world's leading experts when it comes to running great virtual meetings, she's also generous with her knowledge, incredibly patient, and one of the kindest people with whom I have ever had the pleasure to work. I feel honored to have her as my colleague and friend.

Contents

Foreword

Virtual teams are ubiquitous. Years ago, they were confined to specialist groups, but now most meetings have at least one remote attendee. Virtual teams are here to stay. Strong global trends are driving the growth in virtual teams. These include outsourcing, increasingly more global project work, home working and telecommuting, spending cuts, and higher gasoline prices. In recent years, even volcanic ash clouds and the threat of epidemics have played their part. Organizations are focusing on the need to reduce the costs of travel, as well as the time taken traveling, carbon emissions produced, and hassle involved. At the same time, technological advances make it easier and cheaper to collaborate virtually than ever before. Now companies can have teams working around the clock while tapping into a range of global perspectives from across the organization.

Unfortunately though, virtual teams often struggle. Virtual team leaders too often believe that they can apply whatever works for them in the face-to-face world to their virtual teams, and fail to understand what is really needed to make virtual teams work. As a result, team members can become disengaged and frustrated, often silently tuning out. Organizations are only now beginning to realize that specialized skills and competencies are needed to lead effective virtual teams, on top of providing the right combination of virtual collaboration technology tools. Leadership is sorely needed, and that is why I welcome this timely book.

I first came across Nancy Settle-Murphy and her work many years ago when I started to lead virtual teams and came across these challenges myself. It was shortly after the 9/11 disaster. The multinational company where I worked at the time had imposed a complete travel ban worldwide which lasted several months. I was in the middle of running a global program, and needed to get up to speed with virtual working, and quickly. With Nancy's input, I not only survived but thrived in the virtual world. I have gone on to specialize in virtual working on projects and risk management, initially within my former company, and more recently as a consultant. Nancy is not only incredibly effective at helping other people to work virtually, but also models this in her own life. We've worked together on several successful projects over the years and have yet to meet face to face.

In reading this book, I particularly enjoyed the practical tips that can be applied to almost any kind of virtual team. Nancy knows what works and shares them with us in a way that is both easy to read and to apply. She covers the whole gamut from building trust to cross-cultural tripwires, in a way that will work for virtual teams, whatever their makeup.

A couple of years ago I interviewed Nancy as part of the very first Virtual Working Summit and was delighted to share her knowledge. It is a pleasure to write the Foreword of this incredibly useful book, which will equip readers with tips, tools, and techniques they need to become effective leaders of virtual teams. I hope you get as much out of this book as I have.

Dr. Penny Pullan
Host of the Virtual Working Summit and
Director of Making Projects Work Ltd.
Co-author of A Short Guide to Facilitating
Risk Management, *Gower 2011*
Loughborough, England

Acknowledgments

Dozens of people have helped contribute to the contents of this book, both directly and indirectly. Some of them are aware of their contributions, and others may be surprised to learn how much their wisdom and experience have contributed to my knowledge and expertise as a virtual collaboration consultant, and now, a published author!

It was 12 years ago or so that one of my HP clients at the time, Jennifer Grinold (now at Navis Corp.), sent me an urgent e-mail saying, "You know that three-day onsite workshop that you just designed an agenda for? Well, we just got word that no one can travel, so we have to figure out how to achieve the same outcomes, only virtually." "Can't be done," I countered. "You absolutely MUST have the face-to-face interaction, and we can't give up any of that time if we expect to get the work done."

Jennifer pondered my response for a minute, took a deep breath, and said, "Well, we'll just have to find a facilitator who can figure out how to do it."

That was quite a baptism of fire as I suddenly had to learn how to "translate" a design for a productive face-to-face working session into an interactive series of virtual working meetings. And believe me when I say it was not easy. In fact, it took me about quadruple my usual meeting design time to think through the myriad variables we had to factor in to create a series of interlinked virtual meetings that would produce results in a very short time. Thankfully, it was a big success. From that point on, I learned to relish the challenge of designing interactive meetings for virtual participants. (In fact, today I often push back when people insist they need a face-to-face meeting, when they can accomplish the same thing, or better, virtually!)

Over the years, many people have contributed to my learning journey, most of them virtually. Despite the fact that we are 3,000 miles and three hours apart, Julia Young of www.Facilitate.com has been an indispensable guide, teacher, colleague, co-trainer, virtual meeting designer and facilitator, and best of all, friend, over the last decade. No one understands the dynamics and nuances of virtual facilitation like Julia, and few people have her professionalism and patience when it comes to answering my many questions or giving me candid feedback on my meeting designs and course material.

Penny Pullan of Making Projects Work, based in the United Kingdom, is another longtime virtual collaborator with whom I have written articles, brainstormed course ideas, and have relied on for inspiration and support. In particular, we co-wrote an article about the use of visual imagery and metaphor for virtual teams, as well as the virtual meeting checklist in this book.

Closer to home, my colleague and friend Sheryl Lindsell-Roberts (of Sheryl Lindsell-Roberts and Associates), ace business communications trainer, coach, and author of 25 books, has co-written several articles that appear in some way, shape, or form in this book. Our topics have related to intergenerational communications, the use of e-mail, and more effective business communications in general.

Karen Bading of Infrasonics Coaching and Consulting and Charles Feltman of Insight Coaching (both long-distance collaborators from California) are two colleagues with whom I have written articles and designed and delivered virtual workshops related to virtual collaboration, especially in the area of building trust. Robert Whipple, CEO of Leadergrow.com, has co-authored articles with me on the topics of building trust and creating operating principles for new teams.

Two of my go-to gurus over the years have been my colleague Patti Anklam, principal consultant for Net Work from nearby Harvard in Massachusetts, who taught me everything I need to know about social marketing tools and social network analysis, and Michael Sampson, the "Collaboration Guy" from New Zealand, who's given me so many great tips about ways to encourage the successful adoption of virtual collaboration tools. Both Patti and Michael have co-authored a number of articles for my monthly e-zine.

Kate North of e-Work.com has been a great inspiration, with her tireless work in helping organizations around the world to create new virtual workplaces where virtual teams can thrive. (Her company makes some great learning tools for virtual leaders and their teams.) Kate is an active member of New Ways of Working, a global network of organizational innovators who are transforming the workplace by taking an integrated approach to workplace change, combining corporate real estate, human resources, and information technology. (I highly recommend becoming a member if you want to be part of the latest thinking in innovative workplace design.)

Ever since we met more than 15 years ago when I was a change management consultant for his global project team, IT architect Rich Johnston (of UTC's Climate Controls and Security Systems based in Syracuse, New York) and I have co-authored numerous articles, especially regarding recurring challenges related to leading virtual global project teams. He's my role model of an incredibly busy professional who takes the initiative to find new ways to keep learning and hone the new skills needed to succeed as a virtual team leader in an increasingly complex world.

Bart Bolton, facilitator for the Society of Information Management (SIM) Regional Leadership Forum here in New England, has been a colleague and friend for many years. Bart has taught me how and why being a virtual manager is not necessarily the same as being a true virtual leader, and he's co-authored an article with me on that very topic.

This book includes content derived from dozens of e-zine articles I've written over the years, many co-written with other colleagues, fellow consultants, readers, and clients, including

- Kathy Connolly, principal consultant of www.theofficeoutdoors .com, on the topic of galvanizing new virtual teams
- Clint Cuny, CEO of USA at Export Trading Group USA, on the topic of cultural assimilation
- Kristi Ferguson, AVP, Shared Services, Enterprise Real Estate for TD Bank Group, on the topic of coalescing new teams

- Jamie Grettum, senior project lead for The Ken Blanchard Companies, on the topic of why virtual leaders sometimes need to resist the temptation to help
- Hope Kirschner, global marketing manager for an F25 technology company, on the topic of integrating new team members virtually
- Beverly Winkler, senior human resources director in the utility sector, on the topic of recognizing and celebrating great performance from afar

Thanks to the terrific editing and formatting assistance from Peggy Peterson of Peak Editing and Virtual Support in Denver, Colorado, I was able to submit my manuscript ahead of schedule, an amazing feat. Peggy transformed my raw content into a final manuscript, closely following the publisher's specs, with very little guidance needed from me. I could not have done this without her help.

Finally, I want to acknowledge my twin muses, Mayalin and Kiralee, whose busy lives gave me impetus to pursue a profession where I can make a living from home, so I can stay close by to help navigate the rocky road to teenhood (times two). After seeing me on the phone, in front of my computer with headset in place all these years, I think they finally "get" what I do for a living. They are my greatest cheerleaders, most helpful assistants, and always seem to know when a Taylor Swift tune is in order to set the right mood.

Introduction

With so many collaboration tools and technologies that theoretically make it easy for virtual teams to communicate and collaborate, you'd think that leading a well-running, collaborative, geographically dispersed team would get easier. Not really. If anything, the proliferation of new technologies has lulled many of us into thinking that we actually have to think *less* about how we communicate, given how much great technology we have at our disposal. In fact, communicating and collaborating across time, distance, and cultures have never been more complex or difficult.

Let's first take a moment to define virtual teams. For our purposes, we're considering as "virtual" any team that has one or more members working apart from the others. (We don't mean it's not a real team; it's just that members tend not to have much, or any, face-to-face interaction on a regular basis.) In some cases, the leader is the only team member who works physically apart from the others. Or it might be that the leader works in close proximity to some members, and others work remotely. In some virtual teams, the leader and the majority of members work physically together, and just a few work from afar.

People use many different terms to describe a virtual team. Other terms include remote teams, distributed teams, and geographically dispersed teams. Global teams are a particular variant of virtual teams, where members span different cultures and often different time zones. Global project teams are a type of global team where members come together to collaborate on a particular project, and then disband when the project is through.

This book is intended for leaders of all kinds of virtual teams, whatever the nomenclature or configuration, who want to help their team members collaborate more effectively, easily, and enjoyably. We help address a triple-whammy of tough challenges:

- Trusting relationships take far longer for virtual teams to cultivate because there are often few opportunities to create "social capital."
- Cultural differences, languages, and time zones are all considerably harder to traverse in the absence of visual cues and frequent synchronous (same-time) communications.
- A virtual team leader must often influence team members without authority, a difficult prospect in a world of shifting, and often competing, priorities, made yet more difficult when credibility and competency are tougher to establish with those who don't know you well (if at all).

I wrote the hundreds of tips and techniques (in many cases, along with some of my colleagues) you'll find in this book based in part on my own experiences over the last 20 or so years working as a change management consultant to global project teams looking to accelerate adoption of new enterprisewide rollouts. In addition, as a trainer/coach to organizations that need to strengthen their virtual team performance, I've conducted extensive primary and secondary research into what makes great virtual teams tick.

This book can be used as a just-in-time reference guide where you can quickly cull the information you need at the moment. Or you can move from cover to cover, following the sequence of topics laid out here. Or you can start anywhere and skip around in any order that moves you.

Of course, you'll need to consider which tips are applicable to your unique business situation, organizational culture, team composition, sense of urgency, and many other variables. Some tips may not make sense for your team in its current phase, but may be more applicable sometime down the road. We hope that these tips help you keep your virtual team engaged, energized, and aligned through better ways to collaborate and communicate, while actually having some fun along the way.

Please note: I alternate between the use of "he" and "she" throughout to adhere to my practice of writing without bias, while avoiding awkward constructions.

About the Author

A renowned expert in the fields of remote collaboration, global teams, and managing wide-scale organizational change, Nancy Settle-Murphy is a popular author of articles, white papers, e-zines, and booklets. Her articles have appeared in publications such as *The Meeting Professional*, *Mass High Tech*, *IT Executive Journal*, *PM Network*, *Association Management Magazine*, and *Intercom*.

(Photo by David Turton Photography)

Drawing from more than two decades of experience in facilitating the work of global teams, Nancy leads highly productive working sessions designed to efficiently and effectively tap the best thinking of key contributors working across time zones, locations, and cultures. Among her recent clients are Hewlett-Packard, IBM, Shell Oil, Greenpeace International, Medco Health Systems, Partners Healthcare, Unilever, and the Consortium for Energy Efficiency. Nancy is an active member of the International Association of Facilitators, the Virtual Facilitators Linked-In Roundtable, Boston Facilitator's Roundtable, and American Society for Training and Development.

A presenter, trainer, and coach, Nancy parlays her expertise to help answer questions such as

- Under what conditions must we really meet face to face to get work done?
- How can we make remote meetings productive amidst chronic multitasking?
- How do we engage and motivate virtual team members in times of change?
- Can remote teams really cultivate trust when social relationships are not possible?
- What communications methods work best for cross-cultural virtual teams?
- How can we navigate our way around cross-cultural differences?
- What skills are essential for leading virtual teams?

Some of Nancy's most popular speaking topics include Facilitating Cross-Cultural Teams, Building Trust across Borders, Leading Virtual Teams, Planning and Leading Engaging Virtual Meetings, and Jumpstarting a Successful Global Project Team. Nancy is available as a speaker, presenter, writer, and as a source of illuminating content for articles related to virtual teams, global teams, remote collaboration, and managing enterprisewide change.

Nancy is president and founder of Guided Insights (www.guidedinsights.com), a facilitation, training, and communications consulting firm based in Boxborough, Massachusetts, just outside Boston.

1
Unique Challenges
of Virtual Teams
and Their Leaders

Make no mistake: Virtual teams have many advantages over their co-located counterparts. For example, they can make use of a 24/7 workday by parceling out tasks across time zones so their projects never sleep. Team members typically represent a diversity of cultures, skills, perspectives, and capabilities, creating an uncommonly rich resource pool from which all can draw. Plus, virtual team members are more likely to have access to valuable connections and resources they can share for the greater good of the whole team.

People who work on virtual teams often work on large complex virtual projects, which tend to have more visibility, and which, if successful, can help bolster credibility and advance careers of everyone on the team. And, when communications are working well and collaboration is strong, being part of a virtual team can be a rewarding learning experience, and a lot of fun.

1.1 Unique Challenges of Virtual Teams

Trouble is, despite advances in collaboration tools and technology, many virtual teams still struggle when it comes to working in lock step. They want to be able to work more efficiently, effectively, and enthusiastically, but they're not sure how.

The special challenges of virtual teams, especially those that span cultural boundaries, include:

- Large enterprisewide projects tend to be highly complex, with many moving parts, requiring exceptionally well-orchestrated communications and carefully linked activities.

1

- In the absence of face-to-face (FTF) communications, including opportunities to socialize and get to know one another, virtual team members take longer to develop trusting relationships.
- Time zone differences can limit the number and length of real-time conversations, narrowing communication options.
- Cultural and language differences often act as invisible tripwires, making communications frustrating and collaboration difficult.
- Vacation schedules and multiple national holidays mean that many virtual teams have fewer days to get work done.
- Virtual teams tend to operate from an uneven playing field, in terms of proximity to leader or power base, access to resources, sharing of information, ability to socialize, and other factors.
- It's harder for team members to tell whether they're out of alignment about important issues, such as scope, dependencies, accountabilities, and deliverables. And once out of alignment, it takes virtual teams much longer to pull back together.
- Project team members who collaborate virtually, even more than co-located teams, often have multiple reporting relationships, making it hard to assess priorities.
- Giving and getting performance feedback, both across the team and between leader and each member, tends to be less frequent and more awkward.
- It's easier for some people to tune out or renege on commitments when they don't see others on a regular basis, which may jeopardize the deliverables of team members who operate under the assumption that all commitments will be honored.
- Team members have fewer opportunities for the kind of cross-pollination of knowledge and informal learning that co-located teams enjoy during casual conversations.

1.2 Unique Challenges of Virtual Team Leaders

Leaders of virtual teams face many additional challenges of their own, in addition to those affecting the rest of the team. Add to that, because many virtual project team members often work directly for other managers, leaders of virtual project teams have to influence without authority from a distance, with limited opportunities to build relationships that can engage and motivate team members who must constantly juggle multiple priorities.

Other unique challenges of virtual team leaders include

- Establishing credibility and trust between leader and team members, and across the team, takes more work, planning, and time.
- Creating a healthy, open team environment that encourages cooperation and fosters collaboration is hampered with few opportunities for socialization.
- Without the ability to assess the skills, competencies, styles, and preferences of team members, it's harder to match the best people to the given tasks.
- Ensuring that all share the same understanding of team goals, deliverables, roles, accountabilities, and success metrics requires more time and means frequent check-ins.
- Creating a truly level playing field where all team members feel equally valued, respected, and able to contribute fully to the team's success takes a lot of energy, time, and planning, and many may still feel that some are favored over others.
- Developing and agreeing on norms governing vital aspects of communications and collaboration as a team are often skipped (at great peril) due to the time and planning that's really required.
- With fewer opportunities for firsthand observation, determining team members' true performance can be tricky, and as a result, performance feedback may be inaccurate or incomplete. Delivering feedback and performance coaching requires exceptional planning and special skills in the absence of nonverbal cues.
- Detecting when team members have become disengaged, and then offering the appropriate interventions, takes longer without visual cues or frequent contact.
- Maintaining focused productive conversations during virtual team meetings requires special skills that many team leaders don't have. As a result, many team meetings waste time and sap the energy of the team.

1.3 Key Attributes of Successful Virtual Team Leaders

1.3.1 Leadership Skills

- Understands what it takes to establish credibility, and takes deliberate actions to earn it. Knows that credibility is a privilege and not a right.
- Knows how to influence without authority, and goes out of the way to reach and engage each team member, rather than assuming everyone's on board.
- Creates a safe environment where team members know they can surface issues, ask for help, or admit they're struggling without fear of repercussion.
- Finds creative ways to size up skills and strengths to enable effective collaboration.
- Values ability to see problems in different ways. Encourages debates and discussions to get to better ideas and new solutions.

1.3.2 Communication Skills

- Can clearly articulate and communicate a compelling vision to galvanize the team.
- Actively listens so team members feel they are being heard correctly. Knows how and when to paraphrase to ensure shared meaning.
- Discerns communications preferences of each team member, and knows which communication vehicles and styles work best for different team members.
- Detects when team members have become disengaged, disaffected, or otherwise need help getting back on track. Can sense when empathy is needed, even from a distance.
- Knows how to ask the right questions. Understands how certain questions have a way of evoking needed responses.
- Communicates effectively in all respects, including listening, writing, conversational, and persuasive skills.
- Understands that different approaches may be needed with a diverse group, and is aware how own style can affect the quality of communications.

- Values two-way communications and is authentic about the desire for candid feedback, ideas, suggestions, and comments.

1.3.3 Behaviors and Attitudes

- Shows patience and copes well with ambiguity and constant change, and helps others to do so as well.
- Projects enthusiasm and energy. Can be a good cheerleader, both for whole team and for each member.
- Demonstrates sensitivity to cultural, generational, and other differences.

1.3.4 Coaching Skills

- Assesses emotional content of a situation and knows how to dig deeper, resolve, and otherwise address problems.
- Understands motivators for team members, and knows that different members are motivated differently.

1.3.5 Technology Skills

- Understands the range of virtual communication and collaboration tools, and understands how each one can be successfully applied to a given objective.
- Is comfortable using a variety of tools, and helps inspire confidence in others to do the same.
- Knows which, how, and when to use a certain combination of tools to produce the best results.

1.4 Profile of a Successful Virtual Collaborator

Not all project leaders have a chance to hand-pick their own team members. Whether they can choose their own members or have inherited an existing group, virtual team leaders need to know which team members make effective virtual collaborators, and which ones need help getting there. Many who work remotely are poorly suited to make the connections they really need to thrive.

Here are some characteristics that make for a successful virtual collaborator, as well as some attributes that may cause problems for certain team members who have a tough time working remotely. Keep in mind that not all roles require a great deal of collaboration to get work done. Many team roles can, in fact, be performed competently by the "lone wolf" who works independently and remains relatively detached from colleagues. Large, complex virtual teams, however, tend to require that team members remain more linked than ever before, in many different ways.

- **Social butterflies tend to thrive.** The reason: They crave contact with others and are motivated to maintain connections with others, either virtually, through phone, e-mail, or social networking tools, or face to face whenever they can. Introverts who find it painful to stop and chat with an officemate may find it infinitely more difficult to cultivate social connections in a virtual world. Maintaining a deep trusting connection with colleagues is tough for any of us who work virtually, but for someone who is reluctant or introverted, these deep bonds are almost impossible to create and keep up when working from afar.
- **Excellent organizational skills are a harbinger of success.** Virtual workers have to be more self-motivated and disciplined than their office colleagues, because they don't have the luxury of having someone drop in to remind them of an errant deadline or an urgent action. Virtual workers have to set up their own systems for reading, filing, and accessing important content, performing tasks, and reporting progress. Virtual workers also have to follow an established protocol related to the use of file-sharing, e-mail, or social networking tools. Those who are perpetually disorganized or need constant reminders will suffer for their shortcomings even more in a virtual world, with no one there to look over their shoulders.
- **Ability to manage time across many dimensions is a necessity.** Virtual workers must be adept at managing their calendars and syncing up with others, inasmuch as conversations and meetings must be so well orchestrated. They need to be disciplined and realistic about keeping their own calendars,

making sure they build in time for thinking, eating, and moving about throughout the day. Although some cram too many meetings into a single workday, forcing them to work after hours or risk falling behind, others may take too much time "off" for nonwork activities, simply because no one is watching. Those who have a realistic sense of how much time they need to get their work done will be far more productive than those who either can't or won't accurately estimate how much time they need to get work done.

- **They need to be willing and able to use a variety of tech tools with ease.** E-mail and phone as the primary means of team communication have given way to other communication tools. Social networking apps, shared repositories or team portals, instant messaging, texting, web meeting tools, blogs, wikis, telepresence, and more, have become commonplace for virtual teams. Some workers can choose the tool that best meets a particular need, and for others, their organizations have governing principles about the use of certain tools. Regardless, virtual workers have to be conversant about how tools work, and which work best in a given situation, and need to feel comfortable using those tools quickly and easily. Those who are slow to adapt to new communication tools may find themselves being left out of important online conversations or getting only a fraction of the information they need.

- **They need exceptionally sharp listening skills.** People who work virtually miss the vital visual cues that accompany a colleague's disappointment, frustration, elation, or impatience. Virtual workers need to be able to hear verbal cues and read written clues to discern what's really going on for others, much of which often goes unsaid. People from "high context" cultures, where both the context and the words themselves are considered key parts of the overall message, tend to be more successful than those who take another's words simply at face value. Those who don't listen deeply, such as those who chronically multitask during team calls, may never get a sense of the thoughts and feelings that may make or break the success of a virtual team.

- **They know what to communicate, how, and when.** People who know how to organize their thoughts into cogent concise messages have a significant advantage over their colleagues who struggle to put their ideas into writing. Knowing what medium works best for a particular message or a certain situation is a vital skill for a virtual worker who has few chances to make reparations if a message is misunderstood or misinterpreted. Those who insist on e-mails as the default communication mode, for example, may find themselves out of the loop pretty quickly if everyone else is sharing information across a variety of channels.
- **They ignite their own spark.** People who can move ahead without a lot of direction or guidance on a day-to-day basis are far more likely to be successful in a virtual world, where workers must deal with a high degree of ambiguity and the absence of information, sometimes for long periods of time. Those who crave constant feedback or need frequent affirmation will stagnate easily and frequently in a virtual world.

1.5 Summary

To successfully lead virtual teams that consistently deliver superior results, today's leaders need a special set of skills, competencies, and attributes. These can take years, and a great deal of trial and error, to really master. It won't happen overnight, even after reading this book!

What's important is that virtual leaders do a gap analysis between their existing leadership skills and competencies and those outlined in this chapter. Ask your colleagues, peers, manager, or team members to provide input, too. Choose a few skills or attributes to work on first. Create specific goals for yourself, and determine how you'll measure success.

For example, if you suspect that your ability for empathic listening is lacking, try enlisting someone you trust to give you coaching and feedback. Formal training in some areas may be available within your organization or elsewhere. Check in with team members to ask how you're doing, either one-on-one (1:1) or in a team setting.

Sharing your goals with your team lets members know that you're willing to invest a lot of yourself in the success of the team. It can also be a great way to cultivate trust by acknowledging some of your own vulnerabilities and aspirations, which helps create an environment where people can speak the truth, even when they can't see eye to eye.

2

Sizing Up, Onboarding, and Mobilizing Your Virtual Team

On any given virtual team, some members will be brand new and some will be seasoned veterans. Although some team members are full-time, some are part-timers, and others may make cameo appearances only when their expertise is needed.

Even when you have just a few people joining the team for the first time, think of this latest iteration of your virtual team as an opportunity to take stock of the talent, skills, and expertise within the team. Consider anew what connections can be made among team members, and how roles and responsibilities might be parceled out differently.

2.1 Onboarding and Off-Ramping Team Members

By their very nature, many virtual teams have members that frequently come and go. Creating an entry or exit strategy is particularly challenging for teams whose members work virtually. The reason is that unlike their co-located counterparts, virtual team members have fewer opportunities for the kind of informal impromptu conversations by which much vital knowledge is shared and context provided.

Here are some tips to smooth out the transition process:

- **Determine how relatively permanent the arrival or departure of a team member is likely to be.** Plan to spend less time and energy orienting a new member who may be making a one-time cameo appearance, versus one who stands to play a prominent role over time. Likewise, if a team member needs to bow out briefly but will be accessible during this time, you probably needn't expend much energy in planning an exit strategy.

- **Figure out what information a new member needs to launch into productive participation with the team, and how best to provide that knowledge.** If the political landscape is a critical success factor, the best way to describe the likely landmines may be to pick up the phone and use anecdotes to provide needed context. If, on the other hand, the new member needs to get up to speed on the team's progress versus plans, it may be best to point to a shared repository and highlight certain documents that can be read by the new member in advance.
- **Assign a team "buddy" to help ramp-up each new team member.** Try to rotate this responsibility, consciously pairing people who may benefit from each other's skills and experience. Ask the buddy to set aside a certain amount of time during each of the first few weeks to answer questions, provide insight, or give advice. This should be done in person or via phone versus e-mail to allow for more open, direct conversations.
- **Make your team norms and operating principles explicit, and explain how they play out in the day-to-day life of your team.** For example, if a team principle states that each member is responsible for informing others if deliverables are delayed, explain what this means in terms of critical interdependencies, methods of communication, and asking others for help. The more specific you can be about the team norms, the easier it will be to bring a new member on board.
- **When team members depart, make sure to unpack their knowledge and experience before you set them free.** Even though many will promise to make time for the old team once in a new job, competing demands make this promise hard to keep. Think about what content is important for the team to have and where the content is stored. Ask the departing member to write up a few notes, including the status of outstanding deliverables, key contacts, and lessons learned for team members picking up the slack. The more knowledge that can be codified for others to use later, the less costly the loss to the team will be.
- **Plan to have the exiting member interviewed by at least one or two team members well in advance**, giving the team a deeper understanding of the work this person has done, including important relationships, political nuances, and

critical success factors. Even better, have someone on the team shadow the departing team member during important conversations, either virtually or face to face (FTF), to absorb some of the tacit knowledge that may be difficult for the departing team member to describe in words.

- **Negotiate with your departing team members (or their new managers) to borrow a small slice of time** over the coming weeks or months if you have not been able to accurately predict what knowledge is most vital to extract before they leave. Gaining agreement up front will help increase the likelihood that they will make time for you once they've left.

2.2 Assessing Capabilities, Aptitudes, and Preferences of Team Members

When people don't have the opportunity to work together side by side very often, it's hard for them, and for you, to get a feel for their skills, knowledge, communication styles, and preferences. Here are some tips for assessing the skills, preferences, and attitudes of existing team members:

- **Size up skills and strengths to enable effective collaboration:** Virtual leaders need to know how to leverage the strengths, styles, and preferences of all team members so that the whole team benefits. Such an assessment can be tricky with only limited firsthand knowledge or direct observation. Consider using an online tool such as DiSC* or MBTI† to jump start this process. Most such tools are easy and quick for participants to complete, and can yield detailed reports about both individual members as well as the team as whole. Find out if your organization has standardized on a particular type of tool. If in doubt, check with HR or some of your colleagues.
- **Regardless of which tool you use, make sure you can create a team profile that synthesizes all members' profiles** in one

---

* DiSC is a registered trademark of Inscape Publishing, Inc.

† MBTI®, Myers-Briggs, Myers-Briggs Type Indicator, and the MBTI® logo are trademarks or registered trademarks of the MBTI Trust, Inc., in the United States and other countries, and are used under license.

document. This can serve as a handy communications road-map for all team members, and a great tool to help team leaders determine how best to galvanize and mobilize the team during various phases of the project. (Make sure that all team members are on board with sharing their profiles across the team. In rare cases, some people want their profiles to remain private, in which case you'll need to eliminate their profiles.)

- **Think about other ways you can encourage team members to reveal skills, strengths, and other gifts** that may not be obvious from their titles or roles. This can be done in real time, such as during team meetings or via instant messaging (IM), or asynchronously, such as via a shared portal, wiki, blog, or e-mail. For example, you might open a team meeting with a quick tour around the table asking each person to name one special gift they have that will not be obvious, judging from their current role or job title.

- **Assign tasks and leadership roles in ways that take full advantage of the different skills, experiences, and perspectives within your group.** With a long-term project, these assignments and the composition of each group are likely to shift over time. Look for opportunities to shift assignments as a way of keeping people fresh and challenged in positive ways, without disrupting the flow of work.

- **Listen carefully to the tone and content of conversations**, both spoken and written, especially when a crunch is on, because we tend to show our truest selves when we're feeling pressured. What seems to set people off? What behaviors and attitudes are most helpful? Determine which behaviors and attitudes will be most helpful to collaboration, which are impediments, and which need attention, perhaps in the form of coaching.

2.3 Getting up to Speed in a Hurry

Once new people have come on board they'll have to work fast to play catch-up so they can contribute productively as quickly as possible. For virtual teams that are spread across locations and time zones, this game of catch-up can be daunting, both for the newcomers and for those who have to take time to bring their new colleagues up to speed.

Whether your team is brand new, or it's reforming with new members, a virtual team leader needs to find ways to coalesce and galvanize the team quickly, often without many (or any!) face-to-face (FTF) interactions. Here are some actions a virtual team leader can take to create an environment of collaboration and trust in the short term and in the long run:

- **Use FTF as a foundation.** Simply put, without at least some FTF interaction, it's almost impossible for people to develop deep relationships as a truly cohesive team. When people come together in person, they have the opportunity to exchange knowledge and ideas and convey feelings in a way that's tough to do virtually. With the whole team together, you can ensure that everyone is interpreting objectives, goals, roles, and other vital content the same way. Disagreements can be aired more easily and quickly, and mistaken assumptions can be identified and dispelled. FTF meetings help people create the bonds that are needed for people to collaborate virtually down the road. Intermittent small-group FTF meetings can also be critical to get important work done, but nothing can replace bringing the whole team together on a periodic basis to galvanize a team. Yes, it may cost time and money to have everyone travel to one location from time to time, but the risks in not making the investment can be far more costly.

- **Present the big picture right up front.** Without an overall context and framework, new members will have no foundation by which they can make needed changes or suggest improvements. No one is so indispensable that he doesn't have at least some time to get new members up to speed. Appoint a person who has a bird's-eye view of the whole project to orient new members as they come online. (Sending new team members a fusillade of memos is not a fair replacement.)

- **Enable personal insights early on.** When team members rely heavily on e-mail as a primary source of communication, it's tempting to draw conclusions about the meaning of a certain tone or choice of words. Unfortunately, in the absence of any contradictory information, we often make uncharitable assumptions that can fray relationships and erode trust. Provide virtual team members with a way to gain insights into each other's styles, preferences, and behavioral patterns. So, the next time a team member gets an achingly detailed e-mail from a fellow team member, she'll know that this is not meant as an intentional annoyance. Rather, it's just that the sender needs a significant amount of data to process meaning for himself, and is passing that information along to others. For a virtual team, gaining this depth of knowledge about team member preferences may otherwise take months or years.

- **Give everyone a clear sense of how the work of the team fits together.** Do this explicitly, preferably using both words and images to ensure shared understanding. Simply telling people on a conference call or in a conference room, for example, can easily be misunderstood or forgotten. Show people how work intersects and where handoffs are. Go through a couple of scenarios to make sure that the workflow is well understood by everyone. If you lead a large project team, position each team member and each facet of the project with the client as valuable to the overall outcome.

- **Share information openly.** Withholding vital information can impede progress and erode trust. Don't presume to know what information people need to do their jobs. Let them decide. When in doubt, err on the high side. Best to assume that everyone needs access to virtually all project-related information. Try creating a shared portal for everyone to upload access and navigate content easily and quickly, and send alerts to highlight certain information that's particularly vital.

- **Set aside time for real-time team conversations** at least once a week. E-mails, IMs, and the like are not a replacement for the kind of person-to-person exchange that people need to ramp up quickly and feel like part of the team. Establish standards for participation, process, and timing, and make sure that important decisions and actions are communicated in writing to all members afterward, to ensure a shared understanding.

- **Cross-train to encourage easier collaboration.** Discover ways to make it possible for team members to learn others' jobs. This is especially critical for small teams where people have to pinch-hit during peak periods or when team members are absent. Cross-training has other benefits, too. People develop a more holistic perspective of the entire organizational system, and can contribute fresh ideas about opportunities for improvement or growth. Members also gain empathy for others by understanding all they are responsible for, and at the same time, can hone their own skills and knowledge beyond what's called for by their current jobs. Consider who would most benefit by learning each other's roles, mixing together those who come from different functions or organizations. Encourage these small teams to take the best advantage of web tools and shared portals in addition to phone, e-mail, and FTF meetings.

- **Assume that people are trying their hardest.** Avoid making uncharitable assumptions, such as "He's just trying to sabotage our work so we look bad," or "She can't be a team player if she keeps questioning our methods." Chances are, people really are doing their best work, given the circumstances. If people are falling short of expectations, consider how team dynamics are playing a role and work to create better ways of working.

- **Create an environment where it's safe to ask for help.** Don't penalize people who can't meet demands or who ask others to help. If people feel they will be punished for asking for help, you may find them falling behind in silence, with less opportunity for making a mid-course correction later on. Make it acceptable for people to acknowledge they need assistance and make sure they have it, whether it's taking something else off their plate, doling out a new resource, or extending a deadline.

- **Make the time to check in one-on-one (1:1).** When certain members are on the hook to deliver on commitments before others can move ahead, pick up the phone or send a friendly e-mail to see how they're doing and ask how you can help. There's a fine line between micromanaging, which can stop people in their tracks, and making a genuine offer of assistance, which can inspire an energetic performance at the finish line.

As a leader of a fast-moving virtual team, be sensitive to what new members will need to contribute successfully from the start. Never make assumptions about the information you think they need. Show them the big picture and openly share all relevant information. Encourage and respond quickly to questions and provide frequent feedback to assure them they're on the right track. By thoughtfully orienting new members at the outset, the entire team will benefit.

2.4 Assessing How Cultural Differences May Affect Team Dynamics

Even the most seasoned managers occasionally fall into some unexpected and potentially dangerous traps when working with a global virtual team, especially when members are new and relationships have not yet formed. Without benefit of vital nonverbal communications, we tend to ignore or dismiss differences in hopes that having shared goals will be enough to propel everyone ahead at the same time.

Here are some common mistakes many virtual team leaders fall into, along with suggested remedies:

- **Mistake #1:** Assuming that everyone has more or less the same proficiency in writing, reading, and speaking English. Even if your company requires that everyone speak English fluently, some people will be more at ease communicating, whether verbally or through writing, than others.
  Success Strategy: Make sure that you provide multiple communication channels to allow for these differences. For example, if you have a conference call, build in the use of a web conferencing tool to enable more people to participate in different ways with confidence and comfort. In general, allocate at least 30% more time for conference calls to allow for mental translations.
- **Mistake #2:** Arranging meeting times and tasks that will require occasional work on weekends, vacations, and late evenings. Although some Americans may willingly forego personal time for the greater good of their companies, in many other cultures, personal time is sacrosanct.
  Success Strategy: When scheduling work, plan around vacation time and local holidays, rather than asking people to sacrifice private time, and likewise with scheduling team

meetings. If some people have to keep very early or late hours to join calls, rotate meeting times so everyone takes turns being inconvenienced. Also consider using asynchronous means to gather input and ideas from those who may not really need to be on the call at 3 a.m. local time.

- **Mistake #3:** Believing that everyone will be equally willing and able to speak candidly. In some cultures, criticizing others' ideas is considered unacceptably rude, whereas other cultures relish a vigorous debate.

  Success Strategy: Find ways to enable all members to speak their minds safely, even if it means speaking to them 1:1, or offering them an anonymous means of making contributions. Above all, avoid using the "silence is consensus" rule. Otherwise you may imagine you have agreement when in fact you have no idea how certain people really feel or what they think.

- **Mistake #4:** Thinking that all cultures assess trust the same way. Some cultures may place greater value on one's credibility (such as a college degree, related experience, expertise, or seniority), and others may place greater emphasis on reliability (e.g., willingness and ability to follow through on commitments) as a cornerstone of trust.

  Success Strategy: Take the time to discover how different members assess trust, and as a team, consciously create operating principles designed to encourage attitudes and behavior that will do the most to build and cultivate trust.

- **Mistake #5:** Creating a one-size-fits-all team communications plan. Just as individuals may favor certain communication styles, different cultures tend to have different ways of taking in, processing, and sharing information. For example, some cultures require explicit details about their tasks before they can start work, whereas others want only a general framework so they can determine what their tasks should be.

  Success Strategy: Learn enough about all of the cultures represented on your team so you can make some first best guesses about communication preferences. As a team, create some agreed-upon team communication norms that work well for most, if not for all.

- **Mistake #6:** Designing a project plan that requires some members to take on multiple jobs. Before you assume that team members will eagerly volunteer when another member is unable to fulfill stated commitments, verify that in fact each member is able and willing to pinch-hit when needed. Some cultures need roles and tasks to be clearly carved out and feel uncomfortable and at times resentful if they are asked to slide into another role, even temporarily. Other cultures value group harmony over individual achievements and are more likely to jump in when and where needed.

  Success Strategy: Create a team environment where it's OK to say no, to make sure that people don't start over-committing to please you or team members.

- **Mistake #7:** Imagining that everyone has the same definition of "ASAP." Cultures have different notions about time. Americans tend to value immediate gratification and tend to expect that everyone on the team wants to move as quickly as they do. Some other cultures are more deliberate and circumspect before moving ahead and bristle at being rushed. Although some cultures value punctuality, others may regard timeliness as less important.

  Success Strategy: Make sure that everyone regards the milestones and deadlines as realistic and achievable, and be explicit when mapping out deliverables and associated dates. For example, instead of stating that a certain report is due next week, indicate that all reports need to be submitted by 5 p.m. (specifying which time zone if important) on which day.

- **Mistake #8:** Requiring that team decisions be made instantly. Many cultures need to assess input from stakeholders before weighing in. Others prefer making decisions on the fly, often with just partial information. Cultures that value formality may feel disempowered from making decisions without the sanction of their upper management, whereas others may demand an equal vote, whatever their position.

  Success Strategy: Be explicit about how and when decisions will be made, based on whose input, and subject to whose approval. Prepare to build in extra time to allow for some members to conduct the due diligence they need to make decisions if you expect them to follow through later.

The most dangerous mistake people make when leading a cross-cultural team is to assume that we're more alike than we actually are. And with virtual teams, it's much harder to discover if we have inadvertently violated cultural norms or disrespected important values, given the absence of nonverbal cues.

Take the time to understand how different cultures learn, communicate, and collaborate. Stereotypes can be harmful when they are judgmental and inflexible; however, generalizations can be extraordinarily helpful in determining how best to cultivate trust and a sense of teamwork among people of different cultures. Your goal is not to neutralize differences, but to understand how you can make differences strengthen the work of the team.

2.5 Influencing without Authority across Boundaries for Virtual Project Team Leaders

It's tough enough for a virtual team leader to motivate people to stay focused and aligned on shared goals when some or all of the team members are not direct reports. But when the members are geographically dispersed, the leader's challenge is greatly compounded as people constantly struggle to find time for multiple priorities. Oftentimes, the team members whose leader is farthest away in terms of either physical distance or chain of command ends up getting the short shrift.

With everyone carrying a fully loaded "day job" as it is, persuading virtual team members to invest time and energy in your project can be daunting. The first step is to get an honest appraisal of each team member's level of commitment to the project. (After all, just because they agree to attend conference calls doesn't mean they're really present.)

Here are some tips to get you started:

- **Pick up the phone and introduce yourself.** Share your enthusiasm about the project. Let each member know what this project means to the organization and to you personally. Indicate the extent of your expected involvement in terms of time, attention, and activities you will be participating in or leading. Be honest about the challenges you face in pulling the team together, and solicit ideas and guidance for jump-starting the team.

- **Determine whether team members have a real passion for the project.** Do team members see this project as a great learning opportunity or as a necessary evil? Discover the potential benefits of this project from their perspectives, and speak persuasively about the benefits of project membership from your viewpoint. You may not sell them on the benefits at once, but you can start planting the seed.

- **Assess how much time and energy each team member can realistically invest.** Understand the full range of responsibilities team members now carry and how much of their time their current tasks absorb. Ask them candidly how much time they think they can really devote to this project, and over what period of time. For example, some may be able to spend a half-day each week for a month, and others may declare that the best they can do is read e-mail reports to stay informed. Just because someone has been assigned a project doesn't mean he really has time to participate.

- **Set expectations about the kind of help you need from each member and how frequently.** For example, you may require intermittent input and feedback from some team members, whereas others may be required to play a more time-consuming role on an ongoing basis. If you see a significant disconnect between what you feel is required and what this person has to give, be prepared to request a replacement, and notify the team member of your intentions and rationale.

- **Pay attention to communication preferences and styles.** Before you hammer out a team communications plan, take the time to observe how team members communicate most effectively. Do some people communicate more easily in writing? Are some more informative when attending a conference call? Do some members seem peeved by requests for more details, and others are frustrated with "blue sky" conversations? Asking people to go against their grain can heighten resistance and diminish enthusiasm. Be prepared to adjust team communications methods to invite more energetic contributions.

- **Honor the contributions of team members at every opportunity.** Use multiple channels such as phone, e-mail, and web postings to spotlight great ideas or to celebrate the completion of especially important milestones. Make sure that members' managers are kept in the loop. Acknowledge suggestions that lead to positive change. Thank people for ideas even if they cannot be implemented, and be sure people understand why some ideas are adopted and others are not.
- **Invite team members to influence and shape your project goals, strategies, and tactics.** Make sure your request is genuine and not just a token gesture. Acknowledge how this project will benefit by a diversity of perspectives and take the time to ask people what they think. Create an environment where healthy debate and dissenting views are actively encouraged, and build in time for meaningful discussions that will yield richer results.
- **Limit meeting time at first.** Reel them in slowly by taking the time to plan and run a few interesting and productive meetings at the outset, where people have an opportunity to learn and contribute. If you later find you need more frequent meetings to get the work done, you may be pleasantly surprised to find calendars clearing more easily for team sessions.
- **Check in frequently.** Arrange time to meet with members 1:1 at scheduled intervals to find out how they're feeling and what they're thinking. Many people may be reluctant to confess to feeling overwhelmed or slipping behind in a team meeting, especially if everyone feels compelled to play the role of good corporate cheerleader. As team members cultivate trust among each other and for you as their leader, these 1:1 sessions may need to be less frequent.

It's never safe to assume that just because someone has been appointed to a team that she will automatically participate as energetically as you'd like. When you're leading a team of people over whom you have no direct authority, make the up-front investment to earnestly discuss ideas, goals, and challenges. As a result, you're far more likely to convince people to jump on board with both feet. And if they can't make the kind of commitment your project requires, it's better to know up front while there's time to do something about it.

2.6 Summary

Weaving together a whole new team out of members who have different relationships to each other and to the team leader requires exemplary leadership skills and a keen understanding of organizational dynamics. When this newly formed team is virtual, the remote team leader faces an additional set of complex challenges best met by gaining an appreciation for the special dynamics of virtual teams. Everyone can be successful leading a newly formed remote team. It just takes a little more time to learn how and a great deal of creativity.

Before you can expect your virtual team to work at peak velocity and volume, you need to make sure your team is geared up for success. Although this means taking more time and planning than you'd counted on, your up-front investments will be paid many times over with a team where all members are ready, willing, and able to move fast in the same direction, from wherever they work.

3

BUILDING TRUSTING RELATIONSHIPS ACROSS BOUNDARIES

When virtual team leaders name their toughest challenge, one answer always pops to the top of the list: building trust. Or in some cases, it's *rebuilding* trust.

Building trust is hard for any team, but it is especially hard for virtual teams, whose members have few opportunities to interact personally. Virtual teams often evolve around projects, with people coming together and drifting away during different phases. When teams span different cultures, misunderstandings can crop up more frequently with virtual teams, and are much harder to detect, and can be awkward to address. Plus, virtual teams rarely allocate special time for relationship building. So when times are tough, it's almost impossible to drop everything for the kind of heart-to-heart talks that can repair relationships.

Cultivating a culture of trust can take many forms. For starters, virtual team leaders need to find ways to intentionally help team members build relationships that go way beyond the task at hand. As the team leader, it's crucial that you create a level playing field among all team members, regardless of their location or their relationship to you. And it's not enough to treat everyone with the same attention and regard. You have to make people *believe* you are doing so, too. When people perceive that you're playing favorites with some (even when you're bending over backward not to), it's much harder to earn the trust of your entire team.

3.1 Building the Foundations of Trust

Here are some tips to build the foundation:

- **Talk about trust early on.** It can be awkward to discuss trust right up front. (Most teams wait for a team breakdown to bring it up, when it can be way too late.) With a virtual team, it's hard to discern when, how, or why trust has been breached. That's precisely why it's so important to put it on the table at the outset. Encourage people to discuss behavior that tends to build, and break, trust. Guide the team in creating norms that help cultivate trust and minimize opportunities to cause friction. For example, how will team members ask for help or admit when they're behind? What's the best way to deal with frustrations and misunderstandings? Such norms can create guidelines for new members and serve as a checkpoint for all members, should problems arise.

- **Set the tone as the team leader.** The way a team leader interacts with team members on phone calls and in virtual meetings sets the tone for the whole team. This includes tone of voice as well as the warmth with which we say hello. Smiling while talking on the phone can have an impact on others, even if they can't see you. Try it and observe how your tone and tenor changes. Being rested and well prepared for team meetings will result in a positive calm demeanor and a good demonstration of active listening, all of which will have a lasting positive impact on your team.

- **Define trust.** Engage team members in a discussion about trust. Ask why trust is important to them in this group. Not all cultures or people ascribe the same notions of trustworthiness. Ask how they would know if trust had broken down. Ask how they would know if trust were strong. Ask people what constitutes "trust-building" and "trust-busting" behaviors from you, the team leader, and from other team members.

- **Hold each other accountable.** To build trust, all team members need to hold each other accountable to some standards of behavior. If these principles are nothing more than vague intentions or fuzzy "feel good" rules, they won't provide the specificity members need to call each other out in case of a transgression. When leaders permit some members to violate agreed-upon norms, they risk their credibility with team members who expect them to enforce the rules.

- **Reinforce candor.** To foster a culture of trust, the leader needs to ensure people feel safe about voicing their reservations or concerns. The ability of a leader to encourage and reinforce candor lies at the heart of the trust-building process.

- **Anticipate and address stress points.** When people feel pressured to perform, unattractive behaviors can emerge. Without face-to-face (FTF) conversations to smooth ruffled feathers, such behavior can quickly derail even a very strong team. Openly discuss likely stress points and have team members agree how they can best help each other, and themselves, avoid dysfunctional behavior that might result.

- **When in doubt, reveal more rather than less.** Team leaders are often privy to inside information. Err on the side of being more transparent rather than less, providing you don't violate any policies. For example, team leaders might open each call by asking members what rumors they've been hearing, and then address each point with the latest, most accurate information they have. If team members seem reluctant to repeat rumors, try opening an anonymous virtual conference area where team members can pose questions or express concerns, and where team leaders can respond to the team as a whole.

- **Encourage creativity and reasonable risk taking.** Team leaders need to be clear about the type of risks that are encouraged, versus those the organization cannot afford to take. Once ground rules are in place, team leaders can help the team find ways to move creative ideas into action. For example, brainstorming sessions can be set up via phone or virtual conference area where all team members can contribute a volley of ideas to help solve a particular challenge, which can then be vetted and acted upon. Regardless of the outcome, team leaders should congratulate team members for their creative ideas, spawning an innovative, energized environment.

- **Keep an eye out for the small problems.** When virtual team members don't feel comfortable having candid conversations, little annoyances can lead to big problems. Team leaders need to be vigilant about addressing small rifts and immediately bring team members back to the sense of purpose. In some cases, this requires an open conversation with the whole team, and in others, a private conversation may be more appropriate.

- **Give people tangible reasons to make connections on their own.** It's unlikely that a team of 15 or 20 members can build meaningful relationships when conversations are restricted to team calls and e-mail. Assign tasks to team members who need to build trust most urgently. Suggest (or require) that they set aside needed time for important conversations, whether to surface issues, solve problems, brainstorm ideas, or make decisions. During your one-on-one (1:1) sessions with team members, you can learn more about how these sessions are going.

We all know someone we didn't warm up to at first, but as we conversed in the cafeteria, or on a business trip, or in the gym, we developed a close relationship. That personal connection allows us to discover and connect with the full human being. With virtual teams, we have to deliberately create time for the kind of meaningful encounters that lead to trusting relationships. Although this takes no small amount of ingenuity, commitment, and careful planning, the paybacks in happier, more engaged, and productive teams are enormous.

3.2 Creating a Level Playing Field

Here are some tips to give all team members an equal opportunity to fully contribute.

- **Recognize and minimize power differentials, perceived and real.** Be sensitive to the perceptions of remote workers that you may be playing favorites with those closest to you. Ask people on your team for an honest assessment about the extent to which you treat all team members equitably.

You may need to engage someone outside the team to interview members, with anonymity assured, if you suspect
they'll be reluctant to open up to you about your management style. If you find that people feel you're giving more
time and attention to those closest to you (whether you agree
or not), consciously allocate more time to those who work
from a distance.

• **Provide equal access to vital information.** Give everyone,
  regardless of location, the same access to information that
  will benefit the team, at the same time. Resist the temptation
  to share important news with those closest to you. Wait until
  you can reach everyone at the same time. If this is impossible,
  give everyone a heads-up that you need to speak with them,
  and schedule times that are as close as possible to each other
  so the rumor mill doesn't take over before you have a chance
  to share the news.

• **Share power.** Structure the team and activities in such a
  way that power never lies with just one or two, but shifts
  over time. Give everyone a chance to lead something once
  in a while. Grant decision-making authority to those who
  are in the best position to make well-informed decisions,
  versus giving those closest to you more weight in making
  decisions.

• **Share the wealth by stimulating knowledge transfer.** It
  can be tough enough for a centrally located team to share
  members' collective knowledge as a routine part of working
  as a team. A virtual team leader must give extraordinary
  thought to how best to cross-pollinate vital team knowledge, given the limited opportunities for real-time, or synchronous, discussions. Make sure to regularly allocate time
  in team meetings, whether FTF or virtual, for an information exchange, asking members to share what they've
  learned, what they may need help with, or what patterns
  they've noticed. In addition, make it easy to contribute ideas
  or questions in a way that is not time-dependent by using
  team social networking tools, such as blogs and wikis, or a
  team portal where people can push or pull ideas, questions,
  or experience. To jump start knowledge-sharing, consider

some reward or recognition for those who contribute the most ideas, answer the most questions, and so on, within a given timeframe.

- **Maintain the quality of relationships.** Apportion your time evenly among all team members. Make an extra effort to develop relationships with those new to you, or new to the team. If you can't spare time for phone conversations, or if time zone differences make this impossible, try using instant messaging or other ways to reach out to people 1:1 whenever you can. Even a quick, "Hey, how was your birthday party?" or "I really appreciated your help last week," can help make people feel connected, even when they're far away.

- **Make opportunities to provide input.** Seek input from everyone on the team (or explain why input is needed from only some). Give everyone a reasonable chance at offering suggestions, making comments, or providing other input. This means advance planning on your part. Try using asynchronous forms of communication, such as a team portal or some type of online survey, to gather perspectives and ideas.

- **Dole out plum projects evenly.** Give everyone a chance to take on the "best" work. Consider all qualified team members equally for important assignments and interesting tasks, rather than doling out the most coveted projects to those you know best, or those who are closest.

- **Use collaboration technology and tools.** Choose tools that allow all team members equal access from all locations. When some use different software applications for certain team functions, strike a compromise based on what's easiest and most convenient for all to use, when you can. Make sure everyone has enough bandwidth so everyone can participate at the same time, at the same pace.

- **Be aware of cultural and generational differences.** Understand how cultural and generational differences are most likely to affect team collaboration and communication, and do what you can to make sure that no one culture or demographic is at a particular advantage (or disadvantage).

- **Value all contributions equally.** Find ways to recognize and appreciate the perspectives that all bring to the virtual table, regardless of title, tenure, or level. Consider pairing people on certain tasks who can benefit by cross-pollinating knowledge and otherwise learn from each other.
- **Practice bipartisan team leadership.** Chances are you may be physically closer to some team members, or you may be closer to some in other ways due to your work history. Level out the playing field by hosting meetings from different locations, giving people the chance to alternate who participates virtually versus FTF.
- **While in a different location, use the opportunity to walk the halls and check in with team members**, catching up and sharing the latest news. Spend time with people in their usual working environment to give you a new appreciation for the challenges and priorities they wrestle with daily.

3.3 Building Social Capital

Here are some tips to build relationships among team members who have few, if any, opportunities to meet FTF.

- **Use a team kick-off as a time to create social capital, and build from there.** Unlike co-located teams that can bond during FTF kick-off events, virtual teams also need these "getting-to-know-you" sessions. Dedicating an hour-long virtual meeting for a relaxed conversation about family, interests, professional background, and aspirations, for example, can pay big dividends later. Doing this early on gives members more reasons to keep in touch with colleagues, either 1:1 or as a team. Conversations do not, and should not, be all business, all the time.
- **Socialize frequently and celebrate often.** If members are within a reasonable distance, strive to socialize in person as a team or if needed, as two smaller groups. You may have a business reason to get together, such as the celebration of a newly completed milestone or the start of a new budgeting cycle. Or you might have something more personal to celebrate, such as a birthday, service anniversary, or holiday. Even if a

business need is the catalyst for getting together, make sure to allocate some FTF time for relationship-building. Include activities that are both fun and team-focused. For example, for a new team, try asking people to match a member's name to a card with a description of job responsibilities and then assemble an organization chart that shows how everyone fits together as a whole team. Focus on activities that encourage cross-team learning and give people a sense that the whole team is greater than the sum of its parts. Alternate locations to make the commute time more equitable. Ask local employees to host, including creating the agenda.

- **Invite people to reveal a piece of themselves at every opportunity.** Some people dislike "chatting" on team calls, especially when time is of the essence. Others feel they can't really trust another without knowing something about the real person behind the voice. Make it easy and fast for people to reveal a little bit of information about themselves. For example, you can ask people to answer a quick (noninvasive) personal question as they log in or dial in, such as, "What is the title of the last book you've read?" or "Describe what you like best about winter." Little by little, people will develop a deeper sense of the whole person, enabling them to forge connections they may otherwise never have been able to make.

- **Create a way for team members to have "face time."** Even when you can't bring the team together in one room, you can help make them feel like they're together by creating a simple team photo. Ask team members to share a digital photo of themselves, whether a candid shot, family portrait, or professional headshot. Then you can cut and paste the images together to create a team photo, perhaps sitting around an imaginary conference table ready for your next virtual meeting. Add the members' names to their photos and share the collage with the whole team. This simple low-tech tool is a very effective way to keep team members in the mind's eye during virtual meetings and online conversations.

- **Ask everyone to create a "fun fact sheet."** Create a template that team members can easily populate with information that would be both useful to know (such as preferred ways to be contacted, languages spoken, personality profile, etc.) and fun to discover (such as hobbies, favorite vacation spots, family make-up, unusual talents, etc.). People can e-mail their fact sheets to others or, better yet, they can post them in a shared location that is accessible only to team members. Make sure to draw on this information during team calls (e.g., "Jeff, I'll bet you're looking forward to ski season," or "Maria, maybe you can give Juan some local suggestions for his visit to Paris next month"). Bonus: New team members will have an easy way to "meet" everyone on the team right out of the starting gate.

- **Create a space and time where team members can share lessons learned with others.** This sharing can take place in an online conference area, in a wiki or blog, or during team calls. Encourage people to recount lessons that are reflective and revealing, beyond those related to the task at hand. For example, you might ask someone who has recently returned from a client site what they'd do differently next time, and why. Or you can ask each person to call out a quick highlight of the week just ended, or the greatest challenge they face in the week ahead (and why). This can take as little as five or ten minutes on a team call, depending on the number of team members or the nature of the sharing. Alternately, you can call on a couple of different team members each week.

- **Make everyone a star!** It's easy for practically anyone to make and share videos. Ask team members to make videos of themselves in their own work environment (or in their favorite setting) to share with others: maybe during a call, by posting online, attaching in an e-mail, or embedding into a blog. This way, people can get a better feel for each other's environment without having to travel. Consider staging a "best picture" contest for the funniest clip, best acting, most beautiful scenery, cleanest offices, and so on.

- **Reward and recognize achievements out loud.** When your team can't be together FTF, find ways to reward and recognize achievements in other ways. At the very least, use e-mail to notify the whole team of individual or team accomplishments. Better yet, try using videoconferencing for recognizing achievements, so that everyone can feel that he is part of something special. Also think about something tangible you can send to team members (such as team T-shirts, books, flowers, or a savory treat). When people meet virtually most of the time, giving them some sort of three-dimensional team identifier can remind people that they belong to part of a larger whole.

3.4 Summary

To help cultivate trust, build relationships, and maintain a level playing field, successful virtual leaders need to act as the connector for the people on their team. Communications and connections don't happen by chance with virtual teams as they often do with co-located teams who share a corridor, a water cooler, or a cafeteria.

Team leaders need to connect frequently with members themselves, whether 1:1 or in small groups, and create opportunities and suggestions for members to connect with each other. The "connectivity function" is especially important when a new team is forming or new members are coming on board. It's just as vital when the team is going through a rough patch, or when the team needs to be rejuvenated or redirected.

Table 3.1 provides some quick tips for accelerating trust, building social capital, and creating a level playing field. You can use this template and build from there with your own team.

Table 3.1 Quick Reference Guide: Accelerate Trust, Build a Level Playing Field, and Create Social Capital

	ACCELERATE TRUST	
WHAT	WHY	HOW
Create a "dense social network."	• Avoid the out of sight–out of mind syndrome. A virtual team must communicate with one another and with internal stakeholders twice as often as individuals who are co-located or onsite.	• Stay connected with your team. Be hypervigilant in holding ongoing 1:1s; don't cancel them! • Create opportunities for direct reports to work together if possible. • Adopt a mentality of "new month = new connection," where someone on the team introduces the team or a team member to a new internal connection.
Make information easily accessible to all team members at all times.	• Virtual teams rely heavily on the quantity, accuracy, timeliness, and relevance of information.	• Establish principles for information-sharing based on team goals. • Utilize feedback loops with team to gain feedback on what is and what is not working. • Utilize Sharepoint or a central repository for all documents/communications. • Check in frequently to make sure team has information they need to perform and feel connected.
Make it easy to ask for help, acknowledge difficulties. Encourage people to help each other find solutions, offer assistance.	• Create a safe, mutually supportive team environment.	• Model desired behavior by acknowledging your own strengths, concerns, limitations. • Allocate team time for sharing ideas, surfacing issues, problem-solving. • Demonstrate your support as a leader. Rather than giving direction, consider asking: "What can I do to help?" • Support ongoing development plans and efforts for all team members.

(Continued)

Table 3.1 (*Continued*)   Quick Reference Guide: Accelerate Trust, Build a Level Playing Field, and Create Social Capital

BUILD A LEVEL PLAYING FIELD		
WHAT	WHY	HOW
Dole out coveted, visible assignments across the entire team.	• Makes all feel included and equally valued, bolstering engagement and retention.	• Keep track of whom you've tapped for prime assignments. Roll out future assignments equitably.
Give equal air time and stage time.	• Enables everyone to feel equally included with equal time to "shine."	• During meetings, rotate who speaks first. If time is short be sure that the next meeting begins with whomever was missed at the last meeting. • In 1:1 meetings, instead of you leading the conversation, start with, "What would you like to begin with today?" • Help draw out those who have not participated: "I haven't heard from …; what are your thoughts?"
Establish team norms.	• Rules of engagement are critical for virtual teams to ensure alignment around expected behaviors and stem any future misunderstanding. • Your team will know that you are following up with each one of them in the same way and not treating one differently than another. • Establishing team norms helps to create the playing field. How you reinforce the norms (and yes, you will be called upon to do it!) keeps the playing field level or even for all.	• Facilitate a conversation with your team to create norms. • Expected behavior during team calls (arrive early? on time? five minutes late OK? multitasking allowed or not?). • What is expected in terms of communication regarding commitments and due dates? • How do we engage with one another for urgent matters and escalations (get on phone? IM?). • Follow-ups (how frequent, how much, when is it micromanaging?). • What is our commitment to each other regarding turnaround time (TAT) for e-mail and voicemail responses (commit to 24 hours, 48 hours). • Write down team norms, distribute to the team and hold each other accountable if a norm has been broken. Review team norms to any new team members that join.

CREATE SOCIAL CAPITAL

WHAT	WHY	HOW

Quickly get past "just business" mode.	• Team members who are co-located get the chance to look at wedding photos, ask about vacations, comment on the new coffee, and compliment a new hairstyle when they are in the office together. Re-create these moments of social capital for your team.
	
	• Social capital helps you forge deeper bonds across your team, creates a team presence and identity, and demonstrates how you value your team as individuals.
		• Create a team photo. Kids, pets, and loved objects allowed. Request that people post photos in central location. Update periodically.
		• Keep a team birthday, special events calendar.
		   Send cards to celebrants.
		   Celebrate virtually (IM to send wishes, group call, etc.).
		• Use video conferencing as much as possible. Seeing one another is an added bridge or connection. Your effort to hold a video call will be rewarded with double the engagement!
		• Start all your meetings by sharing something personal about yourself. Encourage your team to do the same.
		• Create a team motto or tag line. Carve out time to have fun with this activity. Put it on pens and distribute to your team or use the name when scheduling team meetings.
		• Bridge the geographical divide. Ask an unobtrusive question such as, "Describe what you can see outside your closest window," or "What's next to your computer today?"

Source: Created by Nancy Settle-Murphy of Guided Insights, in collaboration with Mary Rose Wild and Beverly Winkler.

4

BEST PRACTICES
OPERATING PRINCIPLES
FOR VIRTUAL TEAMS

Precious few virtual teams have explicit operating principles, even for aspects of teamwork where the absence of explicit shared values can really trip a team up. Excuses include: "When would we have time to talk this through?" "Everyone pretty much knows how we need to work." "We're too busy." And my favorite, "It's too late to go backward."

And yet, when I ask virtual teams about the toughest communication problems they wrestle with, most of them are resolvable, at least in part, with the creation of shared operating principles. Examples:

- Some people always join our team meetings late, and we waste a lot of time rehashing what we just covered.
- It takes me forever to read through all of the e-mails I get, and more than half of them are totally irrelevant to my work.
- My manager and teammates interrupt my work with a steady stream of IMs all day and get mad when I don't reply right away.
- It takes me forever to find the most current team documents.

It's true that all teams work better with clear operating principles. But virtual teams suffer much more without them. That's because they have so few opportunities to identify, and then successfully address, miscues and missteps that inevitably result when people have different ideas about how they need to work together.

4.1 What Is an Operating Principle?

Some people use the terms "principle" and "norm" interchangeably. For this book, I use "principle" to mean high-level statements that articulate the basic beliefs and values of the organization.

Principles tend to reflect our shared values and perspectives, and provide a stable foundation for a well-running virtual team.

Team norms tend to refer to specific aspects of teamwork, and spring from principles. By developing principles and related team norms, collaboratively through conversation, whether face-to-face (FTF) or virtually, any areas of disagreement, confusion, or conflict can be discussed openly and quickly resolved. Simply put, principles:

- Represent continuity and relative stability in a changing environment
- Serve as a starting point for difficult evaluations and decisions
- Articulate an organization's basic philosophies
- Reinforce and support the goals of the team

For virtual project teams, explicit principles about collaboration and communication can help answer such questions as:

- How responsive must we be to fellow team members, and under what circumstances?
- How often do we really need to meet, and who needs to participate?
- What's the best way to track status?
- How should we use e-mail and instant messaging?
- What's reasonable to expect of other team members during crunch time?

4.2 Principles Development Process for a Virtual Team

Creating meaningful principles does not come easily. Depending on the aspect of teamwork they're addressing, crafting principles that everyone can agree on and accept takes spirited discussion and, frequently, passionate debates. Be realistic about how much time you'll need. The more problems the team is having reaching agreement on certain aspects of teamwork, the more contentious and time-consuming the discussion might be. More benign areas might take less time.

For example, if some on the team feel that other members are not pulling their weight while they routinely sacrifice personal time to get the job done, creating norms about work–life balance is likely to be fraught with emotion. On the flip side, it will probably be relatively painless establishing a new norm pertaining to which version of a particular software application everyone should use.

For a virtual team, it might make more sense to have small groups of people work on drafting team principles for certain aspects for collaboration and communication, such as reaching consensus about change requests. (Creating principles can be a protracted grueling process if you attempt it with more than a handful of people at once.)

You can also try setting up an online conference where people can draft proposed principles for synthesis and discussion as a larger group later on. The key here is to organize the conference in such a way as to make it easy for people to suggest principles without agonizing over the exact words. That will come later. For example, you might pose questions in your online conference area such as these to elicit responses that can be converted into principles and related norms:

- Thinking about our team's use of e-mail, what principles come to mind that might make the use of e-mail more efficient and effective for most team members?
- Considering the demands of this project on our time, what principles would help team members maintain a healthy work–life balance?
- Thinking about the level of engagement during most of our virtual meetings, what principles should meeting planners keep in mind when creating agendas? What norms should participants adhere to if they are to fully participate?

Once your team has agreed on principles, validate and test for understanding and agreement frequently. As you proceed as a team, keep listening, in verbal conversation and online discussions, to get a feel as to whether everyone is still in sync. If you discover that some seem to be going down a different path, state your observations (and encourage others to do so) and probe to find out the reasons.

4.3 Characteristics of Strong Principles

However you choose to guide your team in creating needed principles, make sure that the words have meaning to all. Some principles are so vague that they will do nothing to direct behavior. For example, a norm such as, "We treat each other with respect," really says nothing about how behavior will change. For principles to have meaning, they need to be

- Specific enough to drive behavior
- Stated in the present tense, as though it were true today
- Emphasizes what, not how
- Worded in the positive (vs. a negative: "we do" vs. "we do not")
- Elegant, clear
- Obtainable
- Can be prefaced with "we believe"
- Generates discussion, energy, excitement
- Relatively few in number

Examples:

- We make decisions based on the best information with the fewest number of people.
- Our team communications plan accommodates cultural differences and style preferences.
- All members of our team are free to seek opinions and guidance from other team members.

4.4 Getting to the "So What?" behind Each Principle: Hammering Out Implications

Creating principles is hard work. But coming up with the principle itself is by no means the hardest part. Once a principle is stated, team members need to talk through what, exactly, this means for the team. What norms support this principle? How will we do things differently than we do now as a result? What are the implications? How will we know we're behaving in a way that will actualize this principle?

For example, a principle related to the use of e-mail for one virtual project team might be: "We respond quickly to all urgent e-mail requests from fellow team members." Sounds good as far as it goes, but it doesn't go far enough. Team members then need to agree on specific norms that help support this principle, by answering questions such as:

- What do we mean by "quickly?"
- What constitutes an urgent request?

- Does this apply to all types of requests?
- What about an urgent IM or call?
- How does this change how we work now?

Let's take another example, "Our team communications plan accommodates cultural differences and style preferences." If this team is to uphold this principle, then implications for this team might include:

- We assume that even though all team members speak the same language, team members will have different degrees of proficiency that will affect open communications.
- We rotate meeting times so that people in all time zones take turns being inconvenienced at least once in a while.
- We need to extend meeting time by at least 25% to allow participants to translate in and out of their local language.
- We enable participants to communicate both in writing and via voice in all team meetings.
- We design meetings to balance participation between extroverts and introverts.
- We take the time to discover how cultural differences and communication preferences are likely to affect communication and collaboration in our team, and adjust accordingly.

4.5 Areas for Which Norms Are Especially Vital for Virtual Project Teams

Norms governing certain aspects of collaboration and communication may be more important for virtual teams than those whose members work in close proximity. Here's a partial listing, along with some of the questions that virtual teams can answer through the creation of team norms:

- E-mail
  - When do we use e-mail versus posting documents or sending IMs?
  - What are our conventions around maximum e-mail length?
  - What naming conventions should we consider for subject lines?
  - Do we send attachments in e-mails, or provide pointers? To where?

- What criteria do we use to put some people on the "to" list and others as "cc": or "bcc": (or on no list at all)?
- When do we use "reply all," if ever?
- How many e-mails on a particular topic warrant a call (or an IM)?
- Instant messaging
  - To what extent should team members (and the team leader) be accessible via IM?
  - How do we indicate we're not available, and how seriously should others take our status?
  - To what extent does everyone need to give an accurate indicator of his or her availability?
  - What's a reasonable amount of time to wait for a response via IM?
  - What's a team member's obligation to be available via IM? 24/7, sometimes, or always?
- Document sharing
  - What is our primary method for sharing project documents?
  - Who has access to project documents (on our team, outside our core team)?
  - Who has editing privileges?
  - Who is responsible for alerting team members that new documents or revisions are available?
  - Under what conditions can we use e-mail to send what kind of documents?
- Use of team portal
  - To what degree are team members responsible for uploading and accessing documents from a shared portal?
  - What capabilities will our team use as a default, and which are regarded as options?
  - To what extent do team members need to learn certain functionalities of our portal?
  - How much time do team members reasonably need to review documents posted in our portal prior to a team discussion?
- Responsiveness
  - How quickly do members need to respond to each other's questions or requests for help? Under what circumstances?

- How quickly does the team leader (or designee) need to respond to team members, and what form does the response need to take?
- Decision making
  - Does our team make decisions as a true democracy, or are only some members decision makers at certain times?
  - What criteria will we use to make certain types of decisions?
  - Whose input is required versus whose is requested? Is all input considered equally?
  - To what extent is it important to gather all needed input via multiple channels, especially for those in far-off time zones?
  - Who's responsible for informing team members of important decisions, and in what venue? In what sequence, if not all at same time?

Table 4.1 shows an excerpt of a decision-making matrix one of my clients created for a global project team. For complex project teams where members span several time zones, it's especially important to clearly spell out the decision-making process and related timing. This way, each team member knows which tasks to move ahead with and which ones need to wait. In the absence of such information, costly delays are likely to have a ripple effect on team members around the world.

- Scheduling meetings
  - What are the appropriate duration and frequency of team meetings, based on objectives?
  - What meeting start and end times make sense for members in all time zones?
  - To what extent can team members count on Outlook (or some other shared calendar app) as an accurate reading of a team member's true availability?
- Virtual meetings
  - Do all team members attend certain team meetings, or only some?
  - Under what circumstances can others attend our team meetings?
  - What constitutes full participation (e.g., is multitasking an acceptable practice)?

Table 4.1   Sample Decision-Making Matrix for Virtual Teams

DECISION TYPE	APPROVERS	OTHER PARTICIPANTS BY ROLE	PRIMARY METHOD	SECONDARY METHOD	PREPARATION REQUIRED	COMMUNICATE HOW
Creation of new product/service	VP Finance VP Marketing VP Service VP Manufact.	VP, Sales (influencer) Engineering (influencer) HR (implementer after decision is made)	Face-to-face discussion	Phone meetings, supported by asynchronous (async) conference area	Pro forma projections for sales/profit over next three years	Via weekly conference call to Sales Via e-mail, staff meetings to Marketing
Expanding project scope	Project leader VP, Engineering Funding sponsor	Internal customers (influencer) Sales (influencer)	Weekly project review meetings (phone and web meeting)	E-mail or IM requests (when change requires no additional time, money, or resources)	Detailed accounting of pros, cons, and implications of expanding or not expanding project scope	Via e-mail, status call to all project team members

- What amount of prework is reasonable to expect? What happens if some do it and some don't?
- How do regular team meetings support the team's ability to achieve its goals?
- How are meeting technology choices shaped by our meeting objectives?
- Are all members regarded as co-facilitators, or do we have a clear facilitator for each meeting? Timekeeper? Note-taker? Scribe?
- Whose responsibility is it, if anyone's, to catch people up who don't attend?

4.6 10 Top Norms to Untangle Virtual Teams

Here are 10 best practice norms that can do the most to save time, reduce frustration, and boost productivity of virtual teams. Each norm is followed by a set of specific actions that support each one. For these examples, I touch on virtual meetings, decision-making, the use of e-mail, shared documents, and scheduling—areas for which a lack of explicit norms can cause especially thorny problems for virtual teams.

Although it's not practical or desirable for one organization to simply co-opt another's norms, it can be helpful to see how examples might be modified to fit one's own team. As you read these, keep in mind that a norm that works for one organization may fall flat in another. Use these examples as ideas to jumpstart the creation of norms for your virtual team.

1. **Everyone participates fully in every team meeting he attends.** This means that everyone stays off mute, so we can all hear what's going on, and so people can jump in more easily. Desks, screens, and minds are cleared to focus on the conversation at hand. Multitasking is not acceptable, except for "multitasking on task," such as adding ideas to our virtual conference area or jotting down questions on a shared whiteboard. If you're pulled by competing priorities and can't participate fully, you may need to opt out of a given meeting and catch up on your own later on.

2. **We design meetings to maximize active participation.** This means that we follow the 80/20 rule: our meetings are 80% active and 20% passive. We don't bore meeting participants by showing slides or reviewing documents that can be sent and reviewed before the meeting. Instead, we create our agendas so people can converse on important topics, exchanging ideas, offering suggestions, or seeking guidance. We change activities every five to seven minutes to keep people engaged, and constantly look for new ways to add vitality to each meeting.

3. **We give equal regard to remote and co-located participants.** This means that we incorporate all participants equally in the conversation. We call on remote participants first when going around the table. Onsite participants maintain respectful etiquette, including no sidebar conversations, no putting the speakerphone on mute, no food or beverages near speakers, and saying names before speaking. Whenever possible, we create a truly level playing field by having all participants meet remotely, even when some are able to be onsite together.

4. **Meetings begin and end on time.** This means that the meeting leader starts on time, even if several people are running late. If you arrive late, you are responsible for catching up on your own time; those who came on time are not responsible for repeating what you missed. Meeting leaders need to be realistic about what can reasonably be accomplished within the allotted time, which may mean holding more frequent meetings, longer meetings, having fewer invitees, rescoping objectives, insisting on more prework, or some combination of these things. Meeting leaders end on time, even if all objectives have not been achieved.

5. **We provide sufficient time and adequate information to enable well-informed decision making.** This means that we provide information (e.g., pointers, documents) that people need to digest and reflect on in advance, so people feel prepared to make a logical decision when the time comes. Everyone understands and agrees what criteria will be used to make decisions. We allocate sufficient time for making important decisions, which may mean multiple or extended discussions. We blend

asynchronous (any time) and synchronous (same time) forms of participation to make it easier for everyone to contribute to the discussion, regardless of location or time zone.

6. **Important decisions and their expected impact are communicated to all affected team members at the same time.** This means that all team members, regardless of their proximity and role, are notified at the same time. We resist the temptation to tell those closest to us first. We leave nothing to chance and orchestrate communications carefully, which means giving team members a heads-up about the expected day/time of the team call. We anticipate questions and concerns by those most affected, and come prepared with credible responses.

7. **E-mails are used primarily to inform and alert, versus to distribute documents.** This means that e-mails are kept brief, typically to one screen or less. Links are included for additional information, rather than attachments. Team documents are stored in a shared portal area that everyone can easily access. Document owners are responsible for keeping the portal up to date with the latest revisions of documents. Team members are responsible for accessing documents on their own via the team portal and for setting alerts when newer information is available.

8. **We use "to" and "cc:" e-mail lists with intention.** This means that we define what it means to be on the "to" list versus a "cc:" list. People on the "to" list are required to read the e-mail and take a particular action, which will be clearly stated in the first line of the e-mail. People who are cc'd can opt to read, file away, or delete the e-mail; no action is required. We think carefully about who needs to take action and who should simply be copied, and share our rationale with the whole team in advance to minimize hurt feelings. We welcome feedback from fellow team members if they feel they should be on a different list for future related communications.

9. **We make it easy and fast for team members to respond to e-mails.** This means that we use subject lines that are brief and descriptive. We flag when a request is urgent by denoting a "U" in the subject line. An "urgent" request requires a reply

within no more than four hours, and sometimes sooner. We have a shared understanding of the difference between urgent and merely important. We use bullets or numbers instead of long paragraphs, and embed links for additional information. We confine e-mails to one major topic, to make filing and accessing each e-mail easier later on. When responding to an e-mail request, we revise the subject line so the recipients can read the "short story" without needing to open the e-mail.

10. **Team members rely on shared Outlook calendars to schedule meetings.** This means that time shown as "available" in calendars is fair game for scheduling meetings. We agree in advance whose participation is required, optional, or merely desirable, and indicate as such in meeting requests. We mark off blocks of time we need to get work done. We note vacation time, holidays, and appointments in our calendars. Weekends and local holidays are assumed to be excluded from available time unless otherwise noted. We RSVP to a meeting request ASAP, rather than forcing the meeting organizer to send another round of invitations. If we must decline, we indicate our reasons, so the meeting organizer will know for next time.

4.7 Summary

When working with people you rarely see face-to-face, assume that all operating principles and team norms need to be spelled out and worked through explicitly. Ideally, you'll invest the time in such discussions as a new team springs to life, but in fact, it's only after a team has worked together for at least a little while that the need for certain norms becomes clearer. Caucus the team to discover which aspects of teamwork will improve the most with clear operating norms, and start there, adding more as your team gets the hang of it.

See the Quick Reference Guide (Table 4.2) for creating norms for your virtual team. We provide an example for virtual meetings given the vital role they play in the success of most virtual teams. You can apply these concepts to any area where team norms are needed. What's important is that you agree on specifics as to what a particular norm looks (and sounds) like, and the consequences if the norm is transgressed.

Table 4.2   Quick Reference Guide: Creating Needed Norms for Virtual Teams

CREATING AND ADOPTING NORMS FOR YOUR VIRTUAL TEAM – QUICK LIST

Explicit norms are especially important for a virtual team. Choose one or two as a start, and allocate team time to create a few team norms for each. Make sure to talk through with your team how to sustain each norm, and what the implications are if the norm is violated. Caucus your team to find out where they feel norms are still needed, and set aside discussion time to create them as a team. Periodically check in as to whether norms are still valid or need tweaking.

Team communications as an example:
- Team meetings
- Use of asynchronous conference areas
- Use of e-mail, instant messaging, phone, and texting
- How and where documents will be created, distributed, accessed, and shared
- Work–life balance, scheduling time, being accessible, do not disturb time
- Decision making
- Priority-setting
- Surfacing issues, navigating through conflict

VIRTUAL MEETINGS AS A STARTING POINT

Because virtual meetings are the communications cornerstone of most virtual teams, start there. Here are a few best practices related to virtual meetings. As a team, decide on three or four best practices that your team can adopt as shared norms. Revisit these after your next few team meetings and adjust as needed. Keep adding more norms to your list, branching out to all aspects of teamwork where people feel norms are most needed. Use the templates that follow to begin, and keep building!

Virtual meeting best practices
- Insist on prework by all, as long as it's reasonable
- Everyone off mute to encourage discussion
- Set aside time for check-in or check-out (create social capital at every interaction)
- Design for conversations (80%+ interactivity)
- Be on time; late comers catch up on own time
- Share responsibility for keeping to the agenda
- Take temperature checks when in doubt
- 100% participation, no multitasking (keep track of who is/is not participating)
- Share the air, balance participation
- Keep remote participants visible in our minds' eye
- Rotate responsibilities: facilitator, timekeeper, scribe, host
- At close of meetings, ask participants for feedback on what went well/not well

EXAMPLE FOR VIRTUAL MEETINGS

NORM	HOW WE'LL SUSTAIN	CONSEQUENCES IF BROKEN
Team Meetings		
Keep remote participants visible in our minds' eye	• Go around the room with remote members first • Print out a headshot of all team members and have by phone	• Everyone participates remotely for next few meetings • Remote members set agenda for next few meetings

(Continued)

Table 4.2 (*Continued*)   Quick Reference Guide: Creating Needed Norms for Virtual Teams

NORM	HOW WE'LL SUSTAIN	CONSEQUENCES IF BROKEN
	• Ask remote participants to take turns leading certain meetings • Ask remote team members for candid feedback	
Design for conversations (80%+ interactivity)	• Send out content that can be read and reflected on at least three days ahead of meeting • Minimize presentations or review of documents in all team meetings • Give everyone a few questions they'll come ready to answer	• Participants will opt out, silently • Few people are likely to contribute

WHAT VIRTUAL MEETING BEST PRACTICES CAN YOUR TEAM ADOPT AS NORMS?

NORM	HOW WE'LL SUSTAIN	CONSEQUENCES IF BROKEN
Team Meetings		

Source: Created by Nancy Settle-Murphy of Guided Insights, in collaboration with Mary Rose Wild and Beverly Winkler.

5

COMMUNICATIONS FOR COLLABORATION AND COHESION

The quality of team communications is the single greatest success factor of any project, whether the team works together or apart. And despite this, many teams let their communications methods "evolve" over time, rather than investing time up front to create a thoughtful, well-orchestrated team communications plan. With members geographically dispersed, scattered across time zones and cultures, virtual teams must map out how members will communicate, for what purpose, with whom, and when, right up front. That's because they have so few windows of opportunity for real-time conversations. And if communications go awry, it's far harder and takes considerably longer to make reparations. (Absent a clear plan, it's a matter of *when* communications go awry, rather than *if*.)

Creating a team communications plan is best done face to face, either when the team starts up, or at a critical juncture when the team must work in lockstep to ensure that key deliverables are met. It pays to revise an existing communications plan when a significant number of team members are new, or are rejoining the team, especially if they come from different cultures than most other team members. If face to face is not possible or practical, then allocate sufficient time during your early team calls (or other types of real-time conversations) to do communications planning.

Please note that although we touch on virtual meetings in this chapter, Chapter 10 delves into designing, planning, and running engaging virtual meetings.

5.1 Creating a "Big Picture" Virtual Team Communications Plan

For virtual teams to achieve their greatest potential and take advantage of their diverse knowledge, perspectives, and skills, members must be able to establish a basis for the effective exchange of information, ideas, and capabilities. To do this, the team must create a well-choreographed communications plan that accommodates cultural differences, time zones, preferences, and many different objectives.

Here are some tips:

- **Pay attention to communication preferences and styles.** Observe how team members communicate most effectively. Do some people communicate more easily in writing? Are some more informative when attending a conference call? Do some members seem peeved by requests for more details, whereas others are frustrated with "blue sky" conversations? Asking people to go against their grain can heighten resistance and diminish enthusiasm. Be prepared to adjust team communications methods to invite more energetic contributions.

- **Create a team communications plan with desired performance goals as the starting point.** Rather than simply listing a variety of possible communication vehicles, consider the desired goals of the team and individual members first and work from there. For example, if the team has to develop a new service within 90 days, who needs to know what and when? Does everyone need the same information at the same time? What's the best way to get the information out? Because virtual teams are forced to be more disciplined about communications planning, they frequently outperform co-located teams, whose communications processes tend to be more casual and intermittent.

- **Use multiple ways to connect.** Whatever combination of technologies your team chooses to maintain strong connections—whether phone, web, videoconference, IM, or e-mail—make sure that everyone is comfortable and confident in using the chosen tools. Err on the side of using tools everyone knows, even if the tools have limited functionality.

- **Delineate between the information needs of the core team and the wider group of stakeholders** when mapping out your communication plan. Think carefully before filling up inboxes with information that's irrelevant or off-target. Ensure you are delivering a consistent view of the shared work. Determine who needs to be kept apprised of the latest technical updates to do their jobs, for example, and who might just need a financial summary. Make sure everyone gets the information she needs, which does not necessarily mean the same information.

- **Once people are comfortable communicating, add new capabilities with a clear objective.** For example, move from a tool that allows people to share desktops to one that has an electronic flipchart capability for brainstorming and problem solving. Be explicit as to how certain tools will help the team to achieve certain goals and drive their use. For example, a team portal may be used to enable people to post, access, and edit documents, or weekly conference calls may be the chief venue for surfacing issues. Consider creating an explicit matrix, especially useful for new team members.
- **Consider how team members will reach out to one another, and for what purpose.** For example, is e-mail the preferred way team members ask for help or try to arrange a phone call? Do members pass IMs back and forth during team meetings, or is there a ground rule of total transparency for all communications during team meetings? Whatever your team decides, make sure all members have access to the same capabilities, especially if people work in different groups.
- **Plan face-to-face (FTF) meetings with a purpose.** Carefully think through what you're trying to get out of periodic in-person meetings. Instead of creating an agenda that's full of presentations, progress-reporting, and fact-sharing, consider how you can use at least some of the time for activities and conversations that build trust, nurture relationships, and provide some type of support. Allow members a good balance of social and business interaction. Dinner and a beer after work may be just as important to the mission and success of the team as the formal business content. Good teams work best

when people understand each other well. This understanding is often broadened and deepened in a social context and opens up greater levels of commitment between individuals.

- **Create venues for meeting asynchronously.** Schedule asynchronous team input and activities between meetings to keep team members engaged and aware of progress and process. Avoid long periods of no communication or activity. Use asynchronous web-meeting tools, blogs, wikis, or portals to provide an interactive meeting place. Encourage communication outside the same-time team meetings.

- **Track progress.** Ask participants how they want to share progress updates, including how frequently and in what level of detail. Will all members submit updates at the same time? Will everyone use the same application and format? Do all people have access to others' updates? Agree as to whether all are expected to review updates before joining a team meeting, versus reviewing updates during the call.

- **Make uploading, accessing, viewing, and editing team documents easy and fast.** Create practical ways to share documents and information. If possible, use document archives to avoid multiple copies of documents being circulated at the same time. Use document editing/tracking tools to track changes. Store documents on a shared portal and agree on conventions for naming, review, and updates. Encourage all team members to set alerts for the documents, topics, or authors to which they need to pay special attention.

- **Embrace social networking tools.** First, consider what you want to accomplish by using social networking tools. At the very least, you can create a kind of situational awareness that's not possible via e-mail, phone, or even web meetings. For example, people can discover what other people are working on. Members can be notified when new members join or when people you know connect to other people, join special interest groups, or post information about themselves or their work. Imagine the implications for a team whose members are scattered all over the world, working against a tight timeframe and desperately seeking others who have relevant experience and knowledge to contribute!

- **Provide feedback about the quality of team communications.** Develop norms for providing feedback to each other regarding communication style, quantity, frequency, clarity, and the like. Make sure norms are acceptable across cultures and organizations. This can be accomplished as a team (call, FTF, or a combination), with a survey (anonymous or not), or 1:1.
- **Periodically revisit how well team communications are working,** especially as the nature and intensity of work changes. What's worked at one point in the life of the team or phase of the project may not work later on.

5.2 Virtual Team Communications: Steps to Success

Virtual teams need a plan for effective communications within their own teams, as well as for stakeholders who have a vested interested in the work of the team (see Table 5.1). Such stakeholders might include extended team members, internal or external clients, executive sponsors, training and communications staff, partners, vendors, and others.

Many stakeholders can influence the success of any given project, and not all are obvious. Failure to connect the right people, keeping them involved and informed along the way, can quickly derail the work of a team. Start by defining the different audience segments, their roles, interests, likely communications preferences, and communications objectives, from both their perspectives as well as that of the virtual team. Such a stakeholder analysis can be done face to face or virtually, using a blending of asynchronous and synchronous communications.

Here are some tips for creating a stakeholder communications plan:

- **Calibrate the level and frequency of team communications.** Who needs to know what, and when? How often is too often, or not often enough? Will the team rely more on push or pull communications? To what extent do certain stakeholders take responsibility for selecting how often they will receive information?
- **Establish clear operating norms for team communications and revisit periodically.** In general, anything that is harder to police from afar needs to be clearly stated and agreed to

Table 5.1 Sample Communications Planning Matrix for Virtual Teams

OBJECTIVES	AUDIENCE	MEDIA	TIMING	CONTENT CREATOR	CONTENT APPROVER	OTHER
Share status reports	All team members	• Use Word template • Post onto SharePoint site	• Complete and post at least 48 hr before weekly team mtg.	Each team member	N/A	Keep to no more than two pages
Surface important issues	All team members or those most affected by issues	• Via weekly conference call (if not urgent) • Via e-mail, followed by team wiki if urgent	• Issues that will have an impact on other members' deliverables need to be raised ASAP	Person who identifies the issue	N/A	Err on the side of alerting all team members when in doubt
Track team deliverables at a glance	All team members Management sponsors	• Use of project dashboard posted onto SharePoint site	• Update daily when others' work will be affected • Otherwise, update Fri. end of day (EOD) (local time)	Each team member	N/A	Need to agree on criteria for red, yellow, and green on dashboard

Need quick answer from teammate	Any teammate, subject matter expert (SME), project leader	• IM • E-mail (when not urgent)	• Any time teammates are available to respond	N/A	N/A	Let others know times you are not available to respond to IM or e-mail
Update project plan	All team members	• MS Project	• Access latest rev. via SharePoint	Each team member doing the update	Project mgr. may revise during team meetings	All members set alerts to get instant updates
Inform stakeholders re: progress	Indicate liaisons for each group	• Email is default • Periodic conference calls	• At least 1/wk • More freq. with certain stakeholders	Team comms. mgr. creates reusable content; liaisons modify for each audience	Project leader when comms. need to be especially diplomatic or sensitive	

by all. Other areas for which norms are especially important for virtual teams are handling extraneous topics, punctuality, and preparation and prework. As the team leader, it's vital to state what rules you live by, such as ensuring that meeting times are convenient (or equally inconvenient) for all and actively seeking divergent views to make the best decision.

- **Make sure you have the right people in every conversation, whether asynchronous or synchronous.** Sounds obvious, but it can take a lot of work to decide who really needs to participate in a particular conversation. If people don't really need to be there, they tend to start multitasking or otherwise tuning out. Some people may need to be involved at certain times. Sending out an agenda in advance and spelling out the intended outcomes, along with who will be contributing to each part of the agenda, will take the guesswork out of who needs to be there and the roles they will play.

- **Establish regular checkpoint meetings with the team and keep these times sacrosanct.** If your team spans the globe, consider rotating meeting times so that no one person or region is always inconvenienced.

- **Lead with a clear value proposition.** Create a one-page summary of the objectives, success metrics, risks, assumptions, boundaries, and milestones and refer to this throughout your work together as a team. This supplements detailed project plans to keep the team tuned in to the big picture. Send this out in advance and review key points at the start of the meeting, even if you think everyone's already in sync. Without frequent validation, you may not realize that people are not aligned until it's too late, especially for virtual team leaders who have no nonverbal cues to go by.

- **Intensify communications among those who are heavily dependent on each other for success.** If there are certain people who must rely on each other to get their own work done, suggest (or mandate, in some cases) that people make phone contact at least once every one or two days to check in, surface issues, and report on progress. Especially with projects that are running at a high velocity, team members cannot easily wait for an e-mail that's jammed into a very full inbox or a team meeting that won't take place until the end of the week. As the team leader, you may need to identify these mutual dependencies early on and state your expectations for the nature and frequency of communications.

Disclose more rather than less. Virtual team members are hungry for certain information, especially relating to the state of the project overall, organizational changes, or business conditions that may affect outcomes. Err on the side of giving more such information rather than less, and allocate a reasonable amount of time for people to discuss and absorb the implications. If you sense that team members need to talk, give them the opportunity, along with guidelines to make sure the conversation doesn't spiral out of control. Consider revealing your own thoughts and feelings, especially when changes are made that seem out of your control. See Table 5.2.

5.3 E-mail for Virtual Teams: Using It Wisely and (Probably) Less Often

Despite the advances in collaboration technology tools that open up new ways for virtual teams to communicate across time zones, e-mail continues to be the primary means by which many virtual teams communicate important information. (Of course, the demographics of any given team can affect the extent to which members rely on e-mail vs. IM, text, chat, pop-up videos, and other means of communication.)

If you and your team members are having trouble keeping up with the fusillade of e-mails your team members churn out each day, rethink the conditions under which e-mails are warranted, and agree as a team how you will use e-mail. For example, can the team do without sending attachments, and instead post documents in a shared repository? Is it necessary to ask an important question via e-mail, or is IM or chat more suitable?

Surprisingly few teams have taken the time to create agreed-upon e-mail standards that instill the necessary disciplines to save considerable time and frustration later on. For virtual teams, the absence of well-articulated e-mail practices can have an especially negative impact, given that there are far fewer opportunities to address communication misfires.

Table 5.2 Sample Stakeholder Analysis Matrix for Global Project Team

STAKEHOLDER ANALYSIS: ROLLOUT OF NEW ENTERPRISEWIDE MAIL AND MESSAGING SYSTEM

INFLUENCER	RELATIVE IMPORTANCE (1–5)	ROLE IN THIS PROJECT	DESIRED LEVEL OF PARTICIPATION	"LIKES" (ATTRACTORS)	"DISLIKES" (BARRIERS)	BEST WAYS TO REACH
Business unit leaders	5	• Funder, Sponsor	Visible champion throughout launch and rollout	New system = > productivity Better cross-group collaboration	Fearful of possible downtime, disruptions to business	Through meetings with global project team leaders (face to face and virtual) Send/post brief project updates
Business unit IT leaders	5	• Cheerleader, "salesperson" for new system • Identifies potential issues for project team	Frequent communications to and from business leaders, providing info and support, preparing them for changes			

As more of us scan our e-mails in parallel with other activities (and who doesn't occasionally sneak a peek at e-mail during a team call?), it's especially important to create e-mails with greater intention so we have a better chance to achieve our desired results in less time and with less hassle. Likewise, we need to help other team members be aware of what they can do to increase the likelihood that they focus on our most important issues.

Here are some practical tips, written with my colleague Sheryl Lindsell-Roberts, of Sheryl Lindsell-Roberts and Associates, for creating e-mails that will streamline and strengthen communications among virtual team members:

- **Determine when and how e-mail will be used by the team.** E-mail may be more appropriate for some members than others or for some phases of a team's work. But there may be better options, depending on your objectives and intended audience. Take the time to agree as a team under what conditions e-mail is best, and in what situations another communication channel may work better, such as team portals, IMs, blogs, wikis, and other means. When using portals, make sure people know when new information is posted and provide the necessary links. (By having everyone select his own alerts, you may be able to avoid this step.)
- **Decide on topics for mass distribution versus selective sending.** Avoid the temptation to cover all the bases by routinely sending or copying everyone on every e-mail. For example, determine in advance who needs to be included as a "to" on your status report, who needs to be copied, and who doesn't need to know. Also agree as to whether you'll be using bcc: and in what cases. Check in with team members from time to time to validate your assumptions about their wanting to be included or excluded on e-mails about certain topics. Until you're certain, err on the side of overcommunicating, especially with a new team when relationships are being formed.
- **Establish standards for response time.** Be aware of people's vacation schedules and holidays in other countries. And always remember not everyone is willing to push aside a vacation just because you've marked an e-mail "urgent." If team

members work in a variety of time zones, try setting a standard by which they can respond to e-mail requests by the end of their business days. In this way, colleagues working behind them have what they need at the start of their day. Create conventions to signify urgency in the header so you flag the level of priority. (And make sure that you agree what constitutes urgency. Poor planning on someone else's part does not automatically connote an urgency on your part!)

- **Create a subject line that is clear, concise, and informative.** Type your main message in the subject line. In that way, someone can grasp the gist without having to open your message. Use strong words to grab your readers' attention. For example, if your project is in trouble, instead of writing "Project status" as your subject, write "Project threatened by lack of funding." This will ensure that readers will be motivated to read the text. Filing the message and accessing it later is easier when the subject line reflects the content of the message.

- **Call the readers' attention to actions, issues, and decisions.** The first few lines of an e-mail are critical because they may be the only ones read, especially if your reader accesses e-mail from a mobile device. For example, if adherence to ground rules is important to the success of a meeting, call that out right up front in your meeting e-mail. You may say: "Please arrange your calendars to ensure that we have 60 minutes of your undivided attention for this call. Multitasking will not allow the kind of valuable contributions we need from each of you." Underline key words, use boldface, or highlight in some other way. Headlines may include Action Requested, Next Step, or anything else that is appropriate.

- **Understand the questions your readers will have by asking yourself** *who, what, when, where, why,* **and** *how.* Before you compose your message, consider what questions your readers will need answered. Condense that information into the first few sentences. For example: "First drafts of FY'13 budget plans are due to cost center managers by November 15. All plans must be in Excel format, using the FINPLAN13 budget planning template found in the first entry of our SharePoint directory under the topic 'Budgets and plans.' You can find an example of a completed plan in the document named SAMPLEPLAN 13, listed as the second entry in this same location." By providing all the necessary information up front, you will avoid potential questions (and many unnecessary e-mails) later.

- **Anticipate your readers' likely reaction to defuse negativity.** If your message is likely to be sensitive, contentious, or met with resistance, test it with someone else first (including the subject line). With a virtual team, you have very few opportunities to make amends if you offend or upset someone via e-mail. When delivering negative news, try to offer options or provide a rationale so that people might be more accepting. For example, if you're letting someone know that you cannot complete a report by next Monday, consider mentioning that you can have the first two critical sections by Monday. Always remember: If you're delivering negative news, use the phone or discuss it face to face. Then follow up with e-mail as confirmation, if needed.

- **Eliminate all words and thoughts that don't add value, while being personable and complete.** It's much easier for many of us to spew out as many details as we can think of, leaving our harried readers to extract the hidden kernels. It may require more thought to hone your key message, but ultimately you'll save time by avoiding unnecessary follow-up calls, e-mails, and IMs. When you write an e-mail of substantial length or substance, compose it in your word processor. In this way, you can edit and save the draft for later, rather than feel compelled to hit "send" and accidentally send it before you've had a chance to revise and proofread.

- **Proofread carefully.** Eliminating typos is relatively easy when you use spell check. However, many words have valid spellings that you may have used incorrectly (principal and principle, for example). Also reread for grammar, clarity, flow, and organization. If you question whether you've used a tone that may be offensive, test it with others after you've had a chance to look at it with fresh eyes.

- **Develop cultural sensitivity.** When your team includes people from other countries and cultures, test important messages with people who are fluent in your language and theirs. Make sure your tone is appropriate and your content is clear. Err on the side of formality, especially with new team members who may chafe at a casual salutation or be perplexed by your attempts at humor. Minimize abbreviations and acronyms. (If you must use them, explain them.) Avoid slang and jargon. Use simple vocabulary and conventional syntax. Take the time to check in with people via phone after they've received important e-mails. This ensures that there are no misunderstandings.

Always keep in mind that e-mail is one-way communication. Conversing via e-mail or IM can be time-consuming, distracting, or may result in misinterpretations or misunderstandings. When you use e-mail, make sure your objectives are clear to both you and your reader and that your content reflects those objectives. When you need a real discussion with someone, pick up the phone or schedule a meeting.

5.4 Connecting Virtual Teams through Imagery and Metaphor

Using images and metaphors can work wonders to break the ice, create a shared sense of purpose, and cultivate an environment of real collaboration. But when a team is confined to connecting only through virtual means, the use of visuals as a springboard for meaningful discussion is typically limited. Not because it has to be, but because it takes a lot of thought to figure out how to use imagery when people work miles apart.

Visual concepts can be used to break the ice and connect people in ways that words alone cannot do. Here are some tips for virtual teams, written with my colleague Penny Pullan of Making Projects Work.

- **Build a picture map of the team for all to see.** Grab an image of the state, country, continent, or world where people are based. Insert time zones as needed. Then paste photos of all participants in the appropriate locations. (Ask people to send photos to you or post them in a shared repository.) If you're using a web meeting tool, post this document for downloading. If not, attach the image to your meeting request so all can easily access it as they dial in.
- **Try opening intros.** Consider using an asynchronous web meeting tool to post introductions prior to a conference call.

This saves time on the call and allows people time to consider what they want to share. Try posting questions that evoke an image that will be illuminating for the team. Some examples are to describe your favorite object in your work area and explain why; or tell us what's outside your closest window. Invite members to read the responses before the call so they have time to learn something about each other in advance. On the call, you can have a shorter check-in and reflect on the significance of what each person revealed.

- **Set the stage with a meeting map.** Anchor your meeting with an agreed-upon purpose, agenda, and process. Sound obvious? Many teams gloss over this important step, inevitably leading to longer, less focused meetings. For virtual teams whose members don't have the benefit of a poster or flipchart, try sending a meeting start-up template in advance, and fill in the blanks as a team. If you're short on time, you might send out a completed template subject to revision. You can create your own or use prepared graphical templates. See Figure 5.1.

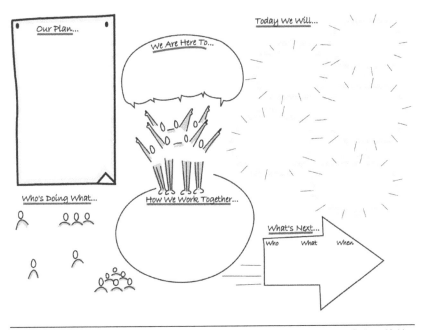

Figure 5.1 Example project meeting start-up graphic. (© Copyright Penny Pullan, Making Projects Work Ltd. Reprinted with permission.)

- **Create a pictorial team charter.** When meeting face to face, many of us find pictures or posters a powerful way to create a shared context. Elicit from team members what images are evoked when they think of the team, the work it's doing, the benefits it will deliver to the organization, and so on. Create a graphical representation of those images, either in electronic or paper form or both, and make it available to all team members. If the goal of the team can be summarized with a single metaphor or image, create a team logo and use it in slides, memos, T-shirts, and the like to create a sense of team identity (this is especially important for virtual teams who don't have many tangible ways of feeling connected).

- **Create a shared view of the present and the future.** Try a web meeting tool to help quickly capture images and adjectives that people have in mind when asking where the organization or team is today versus where it needs to go tomorrow. For example, you might pose questions such as: imagine our group as an animal (or a country); what would we be today? How does this compare to the one we will be six months from now? Once people have entered their responses, invite everyone to view the entire list. Ask for a few volunteers to discuss their responses, and discuss the implications for your team.

- **Use metaphors to get everyone moving in the same direction.** Pick one that's appropriate for this team and its journey, such as white-water rafting, climbing Mount Everest, or flying through the eye of a hurricane. Consider locations and cultures of team members as you choose the best metaphor. Find relevant photos or other images to post during the conversation to evoke the same sense of place for everyone. Get team members talking about what each must do to prepare for this adventure together, what help they need from others, the inherent risks and how to mitigate them, and so on. Capture responses as part of the meeting output, either online or off to the side. "Translate" these responses into real-life implications for your team.

- **Paint pictures with words from the first-person perspective.** Encourage team members to use highly descriptive language, especially when you're limited to an audio connection only.

For example, you might ask, "Imagine you are a typical business manager calling our IT help line. How are you feeling as you dial? Why?" By painting a vivid picture, with each team member imagining she is the focal point, you'll cull more vivid and authentic responses far more quickly than if you said, "Describe the typical client experience when calling your help desk."

- **Choose images carefully.** When you're working with team members who have different native languages, using visual communications is more efficient and effective than using words alone. Tread carefully, however. Make sure that the use of a particular image, whether literal or proverbial, is appropriate and understandable for all team members. Consider the national and business culture, generation, location, and other factors. For example, using an American sports analogy will be confusing and distracting for almost any non-American. Or using mountain-climbing as a team metaphor may be inappropriate for those who hail from flat countries.

- **Use the power of images for instant recall.** After virtual meetings, especially those that were not especially engaging, it can be difficult to remember what happened! Use a mindmap or some other way to visualize the key points. Better still, draw up the mindmap during the meeting with input from all and share it over the web.

- **Keep people focused on a task at hand.** Even if you're not showing a picture, think about ways you can keep participants visually focused while you have them on the call. At the very least, post a team picture on your shared meeting space, as well as an agenda. Think about where and how meeting notes and ideas should be captured in real time. As you construct each meeting agenda, consider how you can keep people visually focused in ways that encourage active engagement from everyone.

When thoughtfully chosen and carefully used, images (whether figurative or literal) can help launch a new virtual team or mobilize one that's derailed faster than words alone. The trick is to think about how we can apply some of the tools and techniques that work well in FTF sessions to engage virtual teams and deepen their understanding of each other and the work they'll be doing together.

5.5 Brainstorming across Borders:
     Stimulating Creative Thinking from Afar

Some people feel strongly that the "best" ideas come from live, in-person "jam" sessions, either because of some superproductive FTF brainstorming sessions they've been part of, or because of some truly terrible attempts at virtual brainstorming, or a combination. Many of today's virtual teams don't have the luxury of FTF meetings.

For those who need to become skilled at adapting brainstorming in a virtual world, here are some tips for productive brainstorming sessions. Please consider both the asynchronous and synchronous collaboration tools you and your team have access to, and adapt these ideas accordingly.

- **Know what problem you're trying to solve.** Much time can be spent on generating a host of ideas that have little bearing on the real problem at hand. For example, one customer service center group initially identified the problem as, "Customers wait on hold too long." The real problem, it turned out, was that customers were bored and annoyed while waiting. One solution was to give waiting customers the option to answer a simple survey while waiting, and give them a modest reward if they do. The results were that Customer Service obtained much-needed client data at relatively little expense, and complaints of long wait times plummeted. The more precisely defined the problem is, the more focused, productive, and quicker the brainstorming session is likely to be.
- **Group similar problems for broader application of great ideas.** With talk time at a premium for virtual teams, and brainstorming sessions apt to be regarded by some as a luxury, try to anticipate some of the problems for which you'll need some creative ideas. Hold a single brainstorming session to generate a flood of related ideas, which can be sorted out later on. For example, a group may tackle these two related issues in one session: minimizing customer "on-hold" time and need to solicit feedback from current clients. Such "bundling" has twin benefits: additional brainstorming sessions might be avoided, and richer results can be realized if some ideas can help solve more than one problem.

- **Consider separate sessions for problem definition, brainstorming, and idea selection.** For most of us, it's difficult to move easily and quickly from the left brain (e.g., the kind of convergent thinking required for problem definition) to the right brain (e.g., the kind of divergent thinking required for brainstorming), and back to the left (for idea selection). Given that meetings among distributed teams are most effective when kept relatively brief, slicing the brainstorming process into a few separate sessions might work best.
- **Make generous use of asynchronous participation.** This is especially important when participants come from different time zones, and when time is at a premium. Post provocative questions in an online conference area designed to elicit creative responses. Encourage people to build on each other's ideas. This way, you can gather many more ideas in a very short time, giving you a great springboard for your team discussion without having to go around the virtual table for ideas.
- **Allocate the appropriate amount of time for each phase.** The general rule of thumb is about 50% or so of the available time should go to problem definition, about 35% to idea generation, and about 10–15% to idea selection. By using certain online capabilities, you can speed up some of this work significantly, especially in the area of brainstorming and evaluating options. Agreeing on problem definition, however, is probably best done through a real-time conversation, which is likely to include some spirited debate.
- **Select the right participants.** Not all people will be appropriate for all phases of the brainstorming process. For example, perhaps a few people who are closest to the topic at hand, as well as the executive sponsor of the initiative, might be involved in defining the problem and articulating associated implications. For a brainstorming session, make sure to include people who display curiosity, imagination, and at least some disinterest in preserving the status quo. Five to seven people is the ideal number. For selecting the best ideas, you probably want a variety of interests, styles, and perspectives represented, as well as someone with the authority to make decisions and present ideas higher up the line. These may or

may not be the same set of people involved in the brainstorming process.

- **Prepare participants for a productive session.** First, make sure your meeting request includes objectives, agenda, sequence of meetings, and the proposed problem(s) to be solved, along with implications (e.g., costs, morale issues, customer defects). Next, consider having participants brainstorm individually ahead of time, to make the best use of meeting time. This can be done online (see above), or by having people think through top-of-mind responses to a few questions in advance (e.g., the three adjectives I would use to describe how people feel when they're trying to use a new software application are _____, _____, and _____). During the actual session, you can do a quick caucus of responses and build from there.

- **Choose the best technology to get the job done.** At the very least, you'll need a phone line for a "same-time, same-place" meeting, as well as some means by which you can capture and record ideas as they come forth. In addition, there are hundreds of tools, methods, and technologies you might consider to foster rapid-fire creative thinking. Some web conferencing services are created specifically for brainstorming, whereas others have features that make pooling and evaluating ideas exceptionally easy and fast. Some allow for anonymity, which may be particularly important in certain situations. Familiarize yourself with the available options. Take a test run to estimate how long your actual session is likely to run, and plan your agenda accordingly.

- **Be prepared to inject some unexpected stimuli.** Many brainstorming experts insist that the best ideas come near the middle or end of a session, when new stimuli are offered after participants claim they have run out of ideas. For example, try keeping a list of nouns or adjectives handy, images you can show, or questions you can ask. One technique that works well for many groups is asking, "What are some of the worst ideas you have to solve this problem?" Not only does this inspire a new spurt of energy, but many of the "worst" ideas can be transformed into some of the best, with just a little tweaking.

- **Make sure you have an agreed-upon way to select the best ideas.** Nothing can deflate an energized group faster than a vague assurance that their ideas have merit and that "something will be done." Be clear at the outset what criteria will be used for evaluating ideas, and who will do the evaluating. Also let people know what decisions will be made as a result, and when and how they will discover the outcome.

Chances are, team members who work virtually are percolating some great ideas that they're eager to share, given the right environment. Today's teams have no shortage of collaboration tools from which to choose. The challenge is how to bring together the best of both worlds: the kind of spirited jam sessions that ensue when a group of people are locked in a room together, combined with a stream of free-flowing ideas that people can offer up virtually, without the constraints of space and time.

5.6 Real-Time Conversations Crucial for Collaboration in a Virtual World

For many virtual teams, communications are all about how über-efficient we can be to get our work done. If we can get away with sending a quick IM or text instead of picking up the phone to get what we need, chances are we will. (A phone call can be risky and protracted. After all, what if the other person also wants something? And what a time-sink it can be if our colleague actually wants a conversation!)

But collaborative relationships are not built on truncated e-mails or cryptic IMs. Like it or not, creating the kind of relationships that foster great collaboration require real conversations that take time, at least occasionally. Making and sustaining connections is much easier when you can catch someone in the hallway or cafeteria for an ad hoc conversation that can build naturally into some type of shared work. In the virtual world, however, you must thoroughly plan every conversation to make sure that participants are convinced that taking any future steps will be worth their while. After all, you have very little time to build an initial connection that will extend beyond that first meeting.

What's important is that we invest time in conversations that are important and meaningful, whether to help us get our work done faster or better, enrich us professionally, expand our thinking, or make remote workers feel as though they're part of something bigger.

These tips, written with my colleague Patti Anklam of Net Work, are designed to help virtual team members focus on planning and facilitating conversations most likely to help cultivate mutually rewarding relationships. We refer here primarily to voice-to-voice conversations, which may entail the use of other tools, such as web-based conferencing or meeting tools. Of course, you can augment those voice conversations with text-based conversations using instant messaging and chat, or asynchronous conversations that may involve e-mail, a bulletin board, or discussion forum of some kind.

- **Knowing what you want to achieve:** If you simply want to establish a connection for follow-up sometime in the future, you may only need one or two conversations. If, however, you want to explore opportunities and take action that will build toward a long-lasting relationship, you'll need at least a few conversations, each one building on the one before. In the virtual world, with very limited opportunities for real-time interaction, we need to be explicit and direct about what we're seeking from the other person, and explain why the connection will be mutually beneficial.

- **Finding a connection that matters:** A conversation designed to discover a basis for relatedness between two people lays the groundwork for forging a deeper connection that can lead to shared action. Tempting though it may be to push for action at the beginning, most people need to relate on some level before taking the next steps. Use IM or e-mail to set up the first conversation and establish objectives. But make sure that your first conversation is voice-to-voice, because so many vital cues are relayed through tones, inflections, and pitch. If the fit seems good, suggest a concrete next step for follow-up. If not, be honest about your intentions so you don't leave the other person hanging. Always acknowledge the others' time and willingness to talk.

- **Paving the way for exploring possibilities:** If you have established a connection that's of mutual interest, the next conversation should allow you to explore potential areas for collaboration. Capitalize on your interest immediately by following up ASAP with specific areas that you feel represent real possibilities, and suggest some days/times that might work for a next conversation. Send relevant documents in advance to help focus the conversation. At the same time, remain open to possibilities you hadn't thought of.

- **Brainstorming ideas:** Plan to brainstorm for possibilities during this next conversation, and then identify opportunities at a later time. This way, you both have a chance to reflect on the possibilities, weighing the pros and cons, before making decisions. Try to book both of these conversations at the same time, one for divergent thinking and one for convergent thinking. Don't leave more than a few days between conversations, or you risk losing momentum. Brainstorming is best done via phone, at a minimum. Web conferencing tools, used either asynchronously or synchronously, can boost output considerably, often within a surprisingly short period of time.

- **Opening up opportunities:** Prepare for the following conversation by sending a summary of the output from your brainstorming session, highlighting a few areas with the most promise. Ask your colleague to do the same. Your goal for the next call is to agree on one or two opportunities that both of you feel are worth additional time and effort. Make sure you're on the same page in terms of intended outcomes, timing, responsibilities, and resources required. Mismatched expectations at this juncture can quickly sour a new relationship. Before you close the call, agree on next steps needed to pursue a particular opportunity.

- **Taking action:** At this point, you may map out suggested action plans in writing, during an ensuing call, or both. Be as detailed as you can in terms of timing, deliverables, roles, metrics, critical success factors, interdependencies, commitments needed, and resources required. Ask that your colleague mark up your document with any changes, or better yet, arrange a call to review your proposed action plan in real-time, jointly editing the document using a web conferencing or meeting tool. Make sure you agree how and when you'll follow up, track progress, surface issues, and make future plans.

- **Checking in and following up:** Although you may be eager to discover what progress your colleague has made, realize that conflicting priorities often intervene, despite the best intentions. If he has missed a deadline, pick up the phone to check in instead of sending a reminder. Likewise, if you can't fulfill a commitment, don't ignore the fact that you're late. Acknowledge your tardiness and explain the reason; be sure to indicate when you will be able to respond. In many cases, checking in can be easily done via e-mail or through another asynchronous means of communication. When issues arise, having a quick conversation can take less time than a barrage of e-mails. Even when no actions are required, pick up the phone or send an e-mail or IM simply to say hello. These social exchanges can act as the best kind of glue to cement long-distance relationships.

- **Closing the conversation:** Not all possibilities lead to real opportunities, and not all opportunities lead to a shared desire to take further action. Be prepared to stop at any point if you both agree, either implicitly or explicitly, that additional time and effort may not be worth it at this point. Agree to what extent it makes sense to stay in touch to keep the connection going, if you both feel it may be fruitful. For example, you might schedule a call for next quarter, include each other on a particular distribution list, or meet face-to-face if the opportunity arises. Or you may both decide that there simply is not enough of a connection to warrant another conversation any time soon. Be sure to acknowledge the other's time and energy, contribution of ideas, and openness to connecting; close with an offer of your own future availability to respond to requests for ideas, connections to others, or problem-solving help.

- **Maintaining the connection:** This person is now in your network, and you should have learned about topics that are top-of-mind for her. You can follow these up by forwarding news items (from e-mail or the web), tossing out ideas, or making introductions to people she may not know.

Forging meaningful new relationships in the virtual world, especially when we're reaching out to those outside our usual working circle, takes considerable planning, effort, and tenacity. To form a valuable new network connection, we need to have real conversations. Persuading each other of the potential value of such conversations is often a tough sell, but one that will be rewarded if you can identify an opportunity and take action in ways that will pay off for both of you.

5.7 Summary

With so few opportunities for real-time interactions, virtual teams cannot leave communications planning to chance. The good news is that most organizations offer a wide variety of virtual collaboration technologies and tools to choose from, so much so that the options can be overwhelming. Start with just a couple of key communication objectives (e.g., status reporting, decision making, problem solving, or issue escalation), and as a team, determine which communications methods and tools, both asynchronous and synchronous, can best be used, and under what circumstances. Create an explicit communications map as you go, expanding it as the team moves forward. Revisit this periodically as a team to validate that it's working well for everyone, and adjust as needed. When new members come on board, take the time to explain your communications plan and be open to incorporating their ideas.

6

MANAGING PERFORMANCE
FROM AFAR

Guiding professional development and managing performance are tough enough for any leader of a fast-moving, super-busy team. For those who lead virtual teams, developing and managing people comes with many different and more complicated challenges, all of which can be overcome with exceptional planning and the cultivation of particular competencies, especially that of active listening.

6.1 Challenges of Virtual Professional Development

From the team members' perspective, the team leader is usually not in their line of sight, making it harder to observe the kind of performance and behavior they seek to emulate if they aspire to rise to the next level. In short, their role models are nowhere to be seen (at least most of the time).

Add to that, what's often missing from virtual teams is the kind of impromptu, informal coaching sessions that many leaders can so easily do with those who work down the hall. Such coaching sessions now must take place through scheduled calls, via e-mail or IM, and occasionally during a relatively rare face-to-face (FTF) meeting. In reality, many of these scheduled meetings are so crammed with day-to-day operational issues that the professional development discussions are often left behind for another day, despite the best of intentions.

Meanwhile, virtual team leaders cannot easily observe the performance of virtual team members. Although they may be able to track members' progress via status reports, team calls, dashboards, and other means, team leaders can't actually observe how their members are working. They can't see what they're struggling with, or what kinds of things they do really well. They can't tell whether members have opportunities for greater efficiencies or more effective

79

ways of working and they can't be sure how well team members interrelate with others. A perennial question many virtual leaders ask is, "How do my team members really spend their time in any given day?"

Team leaders have an incomplete picture as to the competencies, skills, and development needs of their team members. When assessing the performance of team members, virtual leaders often have to make a number of assumptions that need to be somehow validated, which requires an unusual amount of time, thought, and energy. They know that assessing performance according to the number, timing, and quality of deliverables tells only half the story. As a consequence, they may be coaching in the wrong areas, suggesting training that might be irrelevant or unimportant, and most alarming of all, may be missing great opportunities to help their members learn, stretch, and grow professionally.

Team members, on the other hand, often feel handcuffed when they think they're being underused in their current role. They can quickly become frustrated by what they see as limited opportunities to shine and grow concerned that a lack of visibility is stunting their careers.

6.2 Tips for Developing Performance from Afar

In the absence of FTF interactions, virtual leaders need to learn to rely on other cues, and sometimes other people, to gauge the performance and development needs of team members. Here are some tips:

- **Listen for subtleties and nuances.** If you have few opportunities to observe the performance of your team members as they work with clients, peers, or other managers, you need to find ways to discover how others perceive this person's performance. You may be told that "everything's fine" by your employee, but you might get a different story if you asked someone else. Ask gently probing questions if you have any hint that your employee may need some type of intervention from you, and may be afraid to say so.
- **Implement a peer feedback process.** Inasmuch as you can't be with your employees most of the time, you need a way to gather consistent feedback periodically on the performance of

your employees. Determine what method will work best, and be consistent about how you apply this process across your whole team. Let everyone know your intentions, the method you will be using, and how you and each team member will be using the feedback you collect.

- **Be prepared for each development session.** Have a structure in mind for each call, where you might cover the same items each week, such as an update on a personal development plan, discussion of key activities, and areas where an employee may need help or support, information you need to share, your feedback on their performance, and (at least sometimes!) their feedback on yours.

- **Ensure that team members feel accountable for their actions.** Make sure that each team member understands his or her role in achieving performance potential, both short- and long-term. When managers work apart from those they lead, it's especially important that team members take responsibility for their performance and progress, given that you can't monitor their actions. Establish ways to check in periodically, if needed, between your 1:1 calls, so you can provide additional assistance or feedback.

- **Create a real virtual open-door policy.** Open up a "virtual clinic" for all team members where anyone can call in to seek guidance, surface issues, or otherwise get support from you. Set aside the same day or time every week for your virtual open-door office time. (If members span several time zones, you may need to set up two days or times per week.) Such a clinic would be in addition to regular team status calls, when many members may be reluctant to surface tough issues or ask for help. If no one joins, you can use the time to get work done. Refrain from canceling these calls. It may take a while for people to trust this process.

6.3 Challenges of Performance Coaching from Afar

Virtual leaders often have to develop exceptional antennae to even realize there's a need for performance coaching. (By "coaching," I am referring to interventions or other remedies a virtual leader can take to get someone's performance back on track, or to resolve some sort of behavioral problem or issue.) Of course, at times it may be obvious that team members are struggling: when deliverables consistently slip, work quality is noticeably off, or response time suddenly becomes far longer than usual.

But many times, virtual leaders miss the more subtle clues that tell us when some kind of intervention is needed. Are certain people on your team canceling status report meetings? Taking longer to reply to e-mails? Complaining about their work to other team members? Becoming noticeably withdrawn on conference calls? All of these can be signs that people need a catalyst to get back on track.

Assuming for a moment that the virtual leader is clear that an intervention of some kind is needed, making that intervention via performance coaching can be risky and awkward. For starters, team leaders can't read vital nonverbal cues to calibrate what the team member is feeling or thinking during a tough discussion. At the same time, the team member is deprived of seeing the team leader's face to discern the gravity of the situation. You both must be able to decipher dozens of verbal cues quickly, including tone, cadence, lilt, choice of words, use of pauses or silence, throat-clearing, smiling, and sometimes, laughing or crying, all within seconds.

6.4 Tips for Performance Coaching for the Virtual Leader

In addition to holding routine 1:1s and periodic formal performance reviews, virtual team leaders need to conduct many unplanned performance coaching conversations. Although they may be unplanned, they should not be unscheduled.

First off, never surprise someone with an unexpected e-mail or IM that simply says: "Can I have a word with you—*now*?" That's enough to raise anyone's hackles before you even have a chance to speak. Far better to call, e-mail, or IM to agree on a mutually convenient time to speak. Most times you'll want to give them a hint what you want to talk about in as neutral a way possible so they can prepare for the conversation, and assign to it the appropriate level of seriousness (e.g., "I'd like to explore some ideas for improving your conversion rate" or "I've been concerned about your absences and want to make sure everything is OK.")

Here are some tips for planning and leading a virtual performance coaching conversation:

- **Once your conversation is scheduled, before you pick up the phone, find a quiet place to speak,** away from your computer, phone, or other distractions. (Nothing can kill an earnest conversation faster than multitasking.) Have your notes in writing in front of you with any details that may be important, as well as a calendar and a project plan, if appropriate.
- **Listen deeply.** Once you have stated your observations, without judgment, simply be quiet. Allow the other person time to gather his or her thoughts and find the right words, even if it means a minute or two of silence before he speaks. Take notes on a piece of paper and paraphrase every so often to ensure understanding in the absence of visual cues. Carefully ask probing questions for clarification, and only if needed. Refrain from giving advice or opinions during this time. (Make notes if you're in danger of losing good ideas for discussion at another point.)
- **Summarize what you've heard.** Once you're satisfied that you have a good understanding of what's going on, summarize what you've heard as objectively as possible, much as a journalist would report the facts. Pause and seek validation. Ask whether there's anything else that's important for you to discuss before moving on to the next part of the discussion. (Here's where you can ask any additional questions you might have to give you a more complete picture of the other's perspective.)
- **Diagnose the real need.** Perhaps the trickiest part of the whole conversation is knowing how to determine what kind of support a person really needs from you. In some cases, you can come right out and ask. (Be very cautious of your wording and tone here. Asking, "Just what do you want me to do?" is very different from asking, "What would be the most helpful actions I can take on your behalf at this point?")
- **Validate the kind of support you believe he is asking for.** If you sense that he is confused about how you can help or is

reticent to ask, you might try offering specific types of assistance, such as contacting an unresponsive decision maker, delegating a few tasks for now, or pairing him with another team member. Your initial instincts may be to offer help and advice. But tread lightly. If these people doubt their skills or suitability for the task, your offer of help could reinforce those fears. What could be needed is simply time with you acting as a vital sounding board, helping to motivate and focus.

- **Ask permission to give advice or assistance, if you feel it's warranted.** Never assume it's OK. Chances are, your team member will be receptive to your help, but the simple act of asking permission before you offer help can do wonders to help empower your team member and create greater self-sufficiency.

- **Circle back.** Before you end this call, set a time or day with this person to check in to see whether the combination of support and guidance you have offered has made a difference. Also agree on how you will both be kept apprised of actions taken or progress made in the interim. (Once again, use the phone in a quiet distraction-free location for your follow-up meeting to demonstrate how seriously you are taking your commitments to provide her with the needed support.) If you must resort to e-mail, take the time to ask specific questions, referring to notes you've made, versus a terse: "How's it going?"

- **Avoid becoming addicted to adding value at every turn.** Team leaders are genuinely enthusiastic about helping others benefit from their experience. The best team leaders quietly create an environment where others can cultivate competence and confidence without the need for frequent management interventions. Avoid the temptation to make suggestions or provide "constructive feedback" as you listen to others. Instead, show appreciation and encouragement. The more ideas that are allowed to come from others, the less often team members will feel they must validate with you as they move ahead.

- **Practice the power of praise**. Now that your team member is back on track, find opportunities to praise his work, whether it's during a team call, via e-mail, or a phone call. (Of course, you'll want to praise achievements all of your team members have earned.) However, it is especially important to recognize the achievements of those who could use an extra boost to keep them motivated during tough times.

Whether you're speaking with the whole team, or 1:1, acknowledge that everyone will need help (even you). Speak openly about the usual phases a team inevitably goes through and how motivation ebbs and flows naturally for different people at different times. Admitting when you need help (from team members or elsewhere) will give others permission to acknowledge what they may need from you or others.

6.5 You'd Be a Great Virtual Leader if You Could Just Be Quiet: Listening Tips for Virtual Leaders

Virtual leaders must learn to listen for and interpret an enormous amount of information, within seconds, without benefit of body language or eye contact. And we're not just listening for the words that are (or are not) spoken, but also the tone, pauses, inflections, sighs, lilt, laughter, coughing, and perhaps the toughest of all, silence.

Why is it so hard to cultivate and practice superb listening skills? For starters, most business communications training focuses on how to create and present ideas, whether via report-writing, e-mail, or slides. How to listen to others' ideas gets but a footnote. Plus, few management training courses focus on the importance of demonstrating curiosity, crafting the right questions, listening generously, and expressing appreciation of others' ideas. And for many command-and-control managers, taking the time to listen simply impedes progress. ("I don't care what you think. Just do as I tell you.")

Virtual team leaders must act as information hubs for their teams, helping to ferret out, assimilate, synthesize, and share relevant meaningful information across a dispersed team. Knowing how to ask the right questions and listen is the first step. Here are some tips to cultivate better listening for leaders of virtual teams, where some or all members are geographically dispersed:

- **Open yourself to the possibility that other ideas are worth hearing.** It's not enough just to pay lip service by asking, "Any more ideas?" two minutes before your meeting ends. (In fact, pretending to care about other ideas can be far worse than never asking in the first place.) Be honest as to whether your mind is open to assessing and applying new ideas. If you ask for people's opinions, be prepared to do something with them. For example, if you're ready to launch a new marketing campaign tomorrow, don't wait to ask people what they think about it today, unless you're willing to delay the launch as a result of some great new ideas. (And if you do ask for their opinions at this late date, be honest about what you are or are not willing to change as a result.)

- **Don't act like a know-it-all.** Even though you may be the team leader, you don't have to have all the answers. (And besides, you don't even *want* to have all the answers; it's way too much work!) Demonstrate respect for others' perspectives by constantly soliciting opinions and ideas from team members, especially where they're more likely to have fresher information than you in a particular area. Be earnest about your desire to know more and to learn from them. Encourage others on your team to appreciate the diversity of knowledge and experience by reaching out to other team members as well.

- **Know what to ask about.** Take a page from journalists and business consultants, who tend to be naturally curious people and have a knack for posing the kind of open-ended questions that encourage people to speak freely. Think about where your team will most benefit by exchanging or debating ideas, and then craft a series of questions designed to draw out thoughtful responses. Let's say your company is rolling out a new business application. Even though your team can't change the application, they may be able to influence how it's received by others in their organization. So rather than asking general questions about how people feel about the new application (which may be interesting, but may also chew up hours of time), try asking a few questions about how best to launch the application, or which aspects are likely to be most attractive, and to whom.

- **Craft questions that elicit meaningful responses in a short time.** With virtual teams, meetings tend to be brief and attention wanders easily. So it's critical to think about how, exactly, to formulate a question in such a way that people can give a thoughtful response in a relatively short period of time (especially if you're soliciting opinions from an entire team). For example, rather than asking: "So what does everyone think of the new org plan?" try something like, "What's the #1 aspect of this new model your employees will like best? Least?" Or: "Imagine you're about to share the new org model with your team. What's the first question they're likely to ask?" Have several different questions at hand, just in case some don't work in the way you'd hoped.

- **Don't shy away from asking the tough questions.** Given that people can't see your expressions (and you can't see theirs), asking difficult questions can be awkward or downright risky. Even so, it's important that everyone on the team feel comfortable about surfacing sensitive issues or talking through problems when they crop up. Because virtual team members rarely see each other, unless they have a chance to talk things through openly, they'll be left to make assumptions that may be erroneous, or draw conclusions that may be uncharitable. Be careful to pose questions in a way that invites people to be forthcoming as opposed to putting them on the defensive (e.g., "John, can you share the process you used when the VP of Operations called to complain?" vs. "John, what did you do to cause the VP of Operations to get so angry?").

- **Give people a real opportunity to respond.** If you're going to pose a question, let people answer. Think through how long each response might take, and set aside the right time as part of your conversation. Or you can set a time limit ("in two minutes or less, describe … "). This can be especially helpful when you have long-winded participants. When teams span time zones, consider setting up an asynchronous Q&A forum of some type to augment (though not altogether replace) a real-time information-gathering session. Portals, wikis, and blogs can be great for this. For certain sensitive topics, you may want to allow people to provide ideas anonymously.

- **Silence is golden, especially when it's yours.** And this does not mean putting yourself on mute. (In fact, leaders who are perceived to be multitasking during important conversations can lose the trust of their team members astonishingly fast.) Learn how to listen quietly, without waiting to jump in with your ideas. Have a pen and pad handy to jot down key ideas, draw pictures, or do whatever it takes to help you reflect and absorb what others are saying. Periodically interject affirmative comments to show that you're listening, and try to save your questions until the other person reaches a natural pause.

- **Paraphrase to demonstrate active listening.** When you work as part of a virtual team, there's really no way to know for sure if people are listening when you speak. A few monosyllabic responses can't do much to assure people that their thoughts and feelings are really being listened to. Repeat back important points you heard, translating them into your own words, before you build on their ideas or ask a follow-on question. By paraphrasing accurately, you're demonstrating that you not only listened to what the other said, but you've understood their points well enough to restate them. Paraphrasing is an especially important skill when different languages and accents impede shared understanding.

Learning how to listen deeply is an important skill for any type of leader, but for virtual leaders, it's vital. It takes time, practice, and continual feedback from your team to make sure you're getting it right. Start with a few simple steps: next time you're on a call, move away from your screen (and anything else that might distract you), and close your eyes when others speak, so you can really understand what they're saying. You'll be amazed at how well you'll be able to read tones, nuances, and inflections. Keep a list of well-crafted questions handy so you can quickly poll the team at any time. And perhaps most important, circle back to your team to let them know how you've incorporated their ideas, so they'll be energized to contribute more ideas next time.

6.6 Balance Innovation and Expediency for a Supercharged Team

Given the speed and volume of the work we have to get done, many of us have become obsessed with doing more things, faster. For some virtual leaders, it's simply because we can. For others, it's how we've come to be measured. What's getting lost in our single-minded quest for über-efficiency is the relative luxury of idle thought, where we take the time to sow our gray matter with the seeds of half-formed ideas which, with a little bit of nurturing, can spawn big innovations. To sustain competitive advantage, organizations have to innovate constantly. But, thinking creatively takes time and focus, two commodities that are in short supply.

Many companies talk about the importance of innovation; however, many have not set up the conditions for success in a sustainable way. Virtual team leaders can do something about it, by taking responsibility for creating more opportunities for innovative thinking across their teams. Here are a few practical tips, for both individuals and for teams:

- **Give yourself headroom.** When getting yourself from point A to point B, whether driving, walking, or taking the bus, resist the temptation to multitask. Instead of listening to that podcast or hopping onto another call, turn on some relaxing music (Bach, Handel, and Vivaldi are thought to be especially good for getting the creative juices flowing), or just experience silence. Sometimes it's helpful to have a focus for your creative thinking, such as, "What's the best way to motivate my field staff?" Other times, you might do better simply to let your mind unfurl and your thoughts meander. Don't let anyone or anything invade your personal thinking space with distracting clutter and noise.

- **Block out time in your workday.** Creative thinking requires a lot of focused energy, yet few of us actually set aside time for generating new ideas. Our days are just too jam-packed solving problems, making decisions, and measuring results— you know, all of that "critical" stuff that keeps the wheels of progress moving. Mark off brainstorming time in your calendar, and consider it as sacrosanct as any other meeting, if not more.

- **Create the conditions for creative thinking.** Turn off and tune out distractions such as IMs and e-mails. Clear your desktop. Get rid of that stuff you've been wondering where to file, even if it's moving that pile to the floor for now.

Put away that growing to-do list that you never seem to get to. Creating an open space on your desk helps open up space in your mind for fresh thinking. Turn away from your computer to minimize self-induced interruptions. Better yet, find an altogether different venue. Make sure to bring plenty of writing implements (the more colors, the better) and paper, Post-its, or index cards to write on. Bring along a squeeze ball or another tactile object to satisfy your craving to multitask when you've focused on one topic for more than a few minutes.

- **Put brainstorming on the team agenda.** Although it's true that most flashes of brilliance emerge serendipitously, it's also true that if a team does not allocate time for brainstorming or idea-sharing, few opportunities for creative collaboration will present themselves. This is especially true for virtual teams, where real-time team conversations tend to be infrequent and brief, usually adhering to a strict agenda. Try walling off a predictable time each week (e.g., Fridays from 3–4 p.m.) for an "open" brainstorming session for whomever on the team wants to join. Or set aside bigger chunks of time, perhaps less frequently, to generate new ideas to address specific needs (e.g., expanding your donor base or rejuvenating your new onboarding program). If they're not put on the calendar, brainstorming sessions may always be shunted aside in favor of fighting the fire *du jour*.

- **Make it easy to contribute and build on ideas.** Invite people to offer up ideas by using a blending of asynchronous and synchronous participation. Open an online conference area where people can submit their ideas and build on others, whenever it's most convenient. Then, when you're ready to convene in real time, whether virtually or in person, you'll have a rich array of ideas you can use as a springboard for conversation. You can also use this same electronic brainstorming tool during your same-time session, whether participants are onsite or remote. Consider whether participants will feel less inhibited if given the option to submit ideas anonymously.

- **Ask questions that generate energy.** Make sure the problem (or opportunity) is unambiguous and interesting, if not terrifically exciting. For example, instead of asking how to reduce operating costs within manufacturing, solicit ideas about how to get products out the door faster, with less hassle and waste, while maintaining at least the same standards of quality. Ask questions that are broad enough to generate ideas a bit outside the immediate opportunity, yet specific enough to help steer thinking into productive pathways. When setting up an online "think tank," invite representative participants to try out the questions and give you feedback before rolling out to a larger crowd. What seems like a clear question to you might be interpreted in a completely different way by someone else.

- **Keep the ideas flowing, or stem the gusher.** If you have a specific process and end date in mind, call it out clearly, so people will know when to shut off the flow of ideas (e.g., the senior management team will create a short list of projects for further exploration, based upon the following criteria, by February 10, and will announce the final list by March 15). Or, if you want to keep ideas percolating over a longer period of time given the perennial nature of a particular challenge (e.g., how best to reward and retain top talent), keep an online area open, with occasional reminders, alerts, and frequent expressions of gratitude.

- **Bake at least a little brainstorming into every team meeting.** Generating ideas can be a great way to open up a team meeting with excitement and energy, or end a meeting on a high note. Whether you're meeting face to face, virtually, or a combination, set aside at least a sliver of team time to brainstorm new possibilities. Calibrate how much time you'll need based on the richness and complexity of the topic. For example, you may need just five minutes to elicit ideas for a new team logo, but you may need a half-hour to brainstorm ways to train supervisors on a confusing new HR policy. Asking people for their best ideas on a regular basis has a way of making them feel important and inspired.

Although it's tempting to dismiss innovation as someone else's job (preferably someone whose schedule has a lot more vacancies than yours), it behooves each of us, as individuals and as leaders, to carve out opportunities to showcase our best thinking. Generating new ideas is not only fun and energizing, but it's the best way to maintain our competitive edge. The challenge for virtual leaders is to make innovation a priority for your team members by creating the time and space for original thinking, both independently and together.

6.7 Ensuring an Equitable Workload

It's crunch time, and pretty much everyone realizes the need to put aside personal lives for the next few days (or maybe a tad longer) to meet a critical deadline. Trouble is, you discover that although some people are working feverishly to make sure the team meets the deadline, others are adamant that they are not willing to sacrifice their personal lives, again.

During your weekly team status meeting, the tension is palpable. Those who are working long days and nights are sniping at those who won't. Meanwhile, those who put in their eight hours, and not a minute more, seem incredulous that others expect them to sacrifice their personal time. ("I never signed on for this," says one. "I have a life!") Until now, your team has had no explicit norms about addressing workload imbalance. Clearly, it's time to create some before people leap across the virtual table in frustration. Where should a virtual leader (or any leader, for that matter) begin?

Here are some tips I have gleaned from my clients, many of whom struggle with a serious workload as a chronic condition:

- **Set expectations right up front** for all team members as to how many hours are realistically needed from each member at the start of a given week (or at the end of the last one). Ask each person, preferably in a team setting, if they can make that commitment. If not, you may need to either rescope the collective work, or ask some people if they can pick up the slack left by those who can't commit to working the needed number of hours at this time. Reward those who agree to pinch-hit for others. (See last bullet for more.) Of course, anyone joining the team needs a realistic appraisal of the demands of the job before signing on, assuming they have a choice.

- **Reinforce the notion of mutual accountability.** Your team members are working toward a shared goal; therefore, make it clear that everyone is accountable for contributing his fair share. If some team members feel they cannot (or will not) step up to the plate in equal measure during peak times, you need to determine whether this is an exceptional situation or, perhaps, whether the true demands make this job a poor fit.

- **Ask people to make every attempt to address their concerns with other team members privately, as a first step.** Encourage honesty and clarity as to how the behavior of the other person is affecting one's own work. For example, John may not be aware that if he leaves his piece of the project undone as of Friday at 6 p.m., Greta will have to work an extra four hours this weekend to finish up John's work, before starting on her own. Conversely, they may discover by talking it through that if John puts in just a half-hour of extra time on Friday to complete a key task, Greta will be able to get her work done over the weekend, and John can finish his up Monday morning.

- **Recalibrate the overall work of the team to make sure that working excessive hours is not the norm.** Scope creep is inevitable, especially when resources are constrained and pressure to deliver is high. If you find that any team members are continually expected to work overly long hours to meet the demands of the work, as team leader you need to make some decisions to change the dynamic. For example, you may need to renegotiate the workload with your manager or client, or request additional resources, even if only for crunch time. You might also consider how best to reallocate the overall workload, either by shifting tasks, or by shifting the sequence or duration of activities.

- **Give team members a say in determining how best to balance the workload.** Most people like to feel they're pulling their own weight, and few people will knowingly inconvenience their teammates. Set up regular team discussions to review goals, deliverables, workloads, interdependencies, and likely stress points proactively. Encourage team members to decide how best to work together to meet current demands. For example, it might be possible to break up deliverables into smaller chunks, or to skip one task and cycle back later. Which processes can be streamlined? Which work can safely be omitted? Left to their own devices, team members usually do a remarkable job coming up with equitable solutions that restore harmony.

- **Build in a "workload assessment" conversation into every 1:1 meeting.** Some people just aren't willing to speak up in a team setting, especially if they fear they're alone in their struggles. Remind yourself to do this with a list of questions on hand to start off or end each meeting, such as, "To what extent do you feel confident that you can complete X by Friday without having to put in crazy hours? Is there a way another team member can take a piece of your work? Can we figure out how to rescope or reschedule at least part of your work so you can avoid putting in an unreasonable amount of extra time?" or "How important is this work in the scheme of things?" Encourage team members to come to the table with options to explore with you when the workload threatens to become untenable.

- **Separate the "urgent" from the truly important.** Many people classify requests as "urgent" simply because they haven't managed their own time well, or because they have an inflated sense of the relative importance of their work compared to others'. Encourage team members to examine which work is truly important in the scheme of things, and which "urgent" matters can be safely put on the back-burner while attending to the important stuff. Make sure they know you have their backs if they must defend their priorities to those making the requests.

- **Examine work habits.** This is easier said than done when you can't observe day-to-day routines firsthand. If you suspect that someone is putting in extra hours needlessly due to inefficient work habits or a lack of understanding about how to use certain tools, have them walk you through the aspects of their work that seem to be eating up the most time. Brainstorm alternatives together as you go along. Sometimes it just takes talking it through with another person to reveal "aha" ideas for saving time or short-circuiting protracted processes. It may also be the case that this person simply has a problem with time management in general, which can be addressed in a number of different ways.

- **Define what "out-of-office" means to your team.** Be clear about what issues warrant intrusion during regular off-hours and weekends. Under what circumstances can someone be interrupted when taking personal time off, or on vacation, or home sick? Set up a "triage" system within your team, so other team members are able to pinch-hit for those who are unavailable, whether planned or unplanned. Keep a shared calendar that shows major deliverables by person, as well as planned time off.

- **Cross-train and cross-pollinate.** Enable team members to stretch into other areas by giving them the skills, tools, and knowledge they need. Encourage frequent conversations among members most likely to share tasks, and set up periodic "lessons learned" sessions to make cross-pollination and informal cross-training easier.

- **Reward and recognize those who consistently go above and beyond.** For starters, make sure to bestow genuine gratitude, both privately and publicly, for heroic efforts. Reward the hard workers with time off (with no interruptions allowed for work!) from time to time. A caveat: If there are one or two people who consistently stand out for their tireless work, it may be a sign that you need to load-balance the work across the team.

At any given time, some people may be called on to do more, or work longer, than others. Sometimes this inequity is unavoidable, due to the roles people play or a particular phase of a given project. In other cases, a thoughtful allocation of work up front might help prevent workload imbalance. Whatever the reason, when workloads are out of whack, it's time to create some team norms to help apportion work across the team. It can do wonders to restore harmony, if not create perfect equity, and makes for a stronger, happier team.

6.8 Celebrating, Recognizing, and Rewarding Great Performance

It's easy for virtual team members to feel invisible, unnoticed, and unsung, especially those who work far away from the core team. Find ways you can celebrate victories, recognize a job well done, and reward excellent performance for all of your team members, both individually and as a team.

Here are some tips I wrote with Beverly Winkler, a senior human resources director in the utility sector, to help virtual leaders celebrate achievements and recognize remarkable performance for virtual teams.

6.8.1 Creating a Shared Sense of Community

- **Build a sense of community and shared team identity.** For example, logos, T-shirts, and team names all go a long way toward creating a sense of identity that co-located teams feel more easily. Send them out before an important team call, especially one where achievements will be noted. Anything that someone can pick up and touch and see has a way of creating the feeling of "teamness" far more than any virtual depiction of that team.
- **Use videoconferencing to help create the sense of "togetherness."** If you have clusters of team members in a couple of locations, ask them to open up a video stream so they all can see each other as they celebrate.
- **Create a shared space where people can post ideas that worked for them, best practices they have applied, or suggestions that others can use.** Give team members a place where they can share ideas and ahas on their own, freeing you from having to be the conduit every time.
- **Hold a virtual "breakfast bunch" or "lunch bunch" session periodically,** inviting team members to bring a meal and drink and gather around the virtual table to catch up, check in, or just say hello. Adding a video component can help create camaraderie. Such a session would not replace your regularly scheduled team meetings. Rather, it would help to re-create the kind of water cooler or cafeteria conversations that virtual teams have little opportunities for serendipitously.

6.8.2 Recognizing and Appreciating Noteworthy Performance

- **Play to individual preferences.** One person might appreciate a handwritten card; however, another might just as soon receive an unexpected phone call expressing thanks. Consider the preferences, demographics, and culture of each team member as you decide how best to say thanks.
- **Kick off every team meeting with a few minutes of formal and informal kudos to acknowledge recent successes.** When people work virtually, they have few opportunities to share victories, unless special time is carved out. Kudos may come from the leader first, who encourages members to pat each other (and themselves) on the back. Once this kind of self-acknowledgment becomes part of the team culture, members will recount their own successes without prodding, and will more easily acknowledge others' achievements as well. This can also be a great way to end a team meeting on a high note.
- **Show appreciation for contributions, achievements, and sacrifices by making 1:1 contact with each team member.** Send cards, either the paper or virtual kind, or personal e-mails. Or try picking up the phone to say thanks and check in.
- **Acknowledge both team and individual accomplishments.** People who feel their good work goes unnoticed by others will especially appreciate being recognized for it in a public situation, such as a team call or e-mail. Make sure to recognize achievements by everyone, at least from time to time. Use team meetings when you can, and invite senior managers to bear witness to some of your team's more notable achievements. Send e-mails or group IMs to share the good news with others on the team or those outside your immediate team. Even those whose performance is not stellar on a day-to-day basis appreciate kudos from time to time.
- **Send something from the heart.** This might take the form of a handwritten card, a meaningful book, a fun desktop object, or even a basket of fruit or flowers. Few people take the time to send real cards or specially chosen gifts these days, and team members will appreciate that you went the

extra mile for them to do it. These important messages really let individuals know they matter to you and the business. An added benefit is that family and friends of a particularly hard-working team member may be impressed by the gesture, and may just be a little more understanding the next time a family event has to be missed or postponed to meet a work deadline.

- **Make it official.** When people work virtually, they appreciate having something tangible in their hands to remind them that they belong to a "real" team that's worthy of being recognized. You can find templates online, or right in your slide-making or word-processing software, that make it easy to create and print custom-made certificates. The certificate can denote a team achievement (e.g., Excellence Award for Exceeding Sales Goals in Q3), or it can reflect an individual achievement, such as best sense of humor in the face of adversity or most improved performance. Mail the certificate in an envelope so the individual can open it when the achievement is announced during the team meeting.

- **Call your team members on the phone.** Sounds simple. But in today's techy age, we all IM, text, and e-mail. The sound of your voice will add enormous credibility and provide a specific message that's meant especially for this person.

- **In addition to using phone, e-mail, IM, or team meetings to say thanks, also try web postings** to spotlight great ideas or to celebrate the completion of especially important milestones. Make sure that members' managers are kept in the loop.

- **Shout it from the virtual rooftops.** Of course, thoughtful e-mails that recognize great performance are also much appreciated, especially if the message reflects that the sender clearly took time and thought to compose it. Copying a wider audience, such as fellow team members and other leaders, can amplify the positive effect.

6.8.3 Planning and Running Virtual Celebrations

Virtual leaders know that, perhaps even more than co-located teams, virtual team members need opportunities to celebrate achievements

People who work virtually need constant reinforcement that they matter: to each other, to their leaders, and to the overall organization. Although it takes more planning and creativity to celebrate achievements of individuals and virtual teams alike, the payoff is great. Remember, celebrations are the moments when team members feel like they belong to something bigger outside their virtual office. And the more team members feel their contributions are appreciated, the more engaged, energized, and motivated they are likely to be. Plus, you'll have more fun as the leader too!

6.9 Summary

All aspects of managing performance, of both individuals and of your entire team, need to be recast when your team is virtual. Some team members will require more of your time and attention than others, the degree of which might ebb and flow over the life of a team or the span of a particular project.

Be prepared to allocate more time than you think you'll need for these important conversations, especially when the team is new, or when members have recently joined. By setting aside the time these conversations deserve, you'll give team members the confidence and competence they need to become self-sufficient in their current roles, and infinitely more capable of stretching to reach even higher levels of performance. The extra payback for you as team leader is that ultimately, this will mean they'll need far less of your time to make decisions or take action.

The sample and template shown in Table 6.1 gives you an idea of what a performance action plan might look like for each team member. You might have different ideas for the "aspects that need addressing" for your team, or for a particular team member.

What's important is that you have a way to keep track of action plans that you or your team member will take, by target date, along with the expected outcome. The expected outcome might be quantifiable (John will respond to e-mails within 48 hours), or qualitative (Maria and her team will feel more appreciated and inspired to take their performance to the next level).

Whether you use this template or one of your own design, make sure you have a consistent place where you take careful notes and have them ready for each 1:1 meeting.

Table 6.1 Managing Performance from Afar: Quick Reference Guide

MANAGING PERFORMANCE—ACTION PLAN BY TEAM MEMBER—SAMPLE

TEAM MEMBER	ASPECT THAT NEEDS ADDRESSING			HOW/ACTION PLAN	WHEN	EXPECTED OUTCOME
	ENGAGE AND MOTIVATE	DELIVER PERFORMANCE FEEDBACK	SUSTAIN GREAT PERFORMANCE			
Sample: John Doe		X		• Leverage 1:1 Meeting … Preplan by framing a feedback statement that focuses on behavior and impact on team (e.g., lack of responsiveness)	August 8	• Reach mutual agreement to stay with team norms (respond to e-mails within 48 hours) Revisit at next 1:1 meeting
	X			• Assign as team meeting facilitator for month of Sept., with my support – gauge interest at mid-Aug. 1:1	August 15	• John will feel more energized at team meetings and will hone his skills in facilitation and time management
Sample: Maria Johnson			X	• Acknowledge the performance of her team over the last month at upcoming team meeting. Send personal thank you card for Maria to share with her team.	Team meeting July 28 Send card July 22 or sooner	• Maria and her team will know how much we appreciate their amazing performance and will be inspired to take it to the next level

MANAGING PERFORMANCE—ACTION PLAN BY TEAM MEMBER—TEMPLATE

TEAM MEMBER	ASPECT THAT NEEDS ADDRESSING			HOW/ACTION PLAN	WHEN	EXPECTED OUTCOME
	ENGAGE AND MOTIVATE	DELIVER PERFORMANCE FEEDBACK	SUSTAIN GREAT PERFORMANCE			

Source: Created by Nancy Settle-Murphy of Guided Insights, in collaboration with Mary Rose Wild and Beverly Winkler.

7

NAVIGATING ACROSS CULTURES, TIME ZONES, AND THE GENERATIONAL DIVIDE

Most virtual teams that work within large global organizations tend to span cultures, generations, and more often than not, multiple time zones. And even when most team members work in close proximity to each other, chances are many cultures and generations will be part of the mix.

Navigating through the trip wires of multiple cultures can be tough even when team members have the ability to read nonverbal cues such as body posture, gestures, and facial expressions to help decode the meaning of certain words. But absent the ability to correlate visual cues to written or verbal expressions, with few opportunities for real-time conversations, members of global teams often struggle to identify and address barriers that get in the way of successful communication and collaboration.

There are many reasons. For starters, it can be awkward and uncomfortable to discuss cultural differences openly. Many leaders have little experience in discussing cultural differences, and may be afraid that they will be perceived as someone who unfairly stereotypes members of other cultures. In addition, some leaders really don't understand how cultural differences are affecting their particular team, either because they have not taken the time to learn about the differences, or they don't know where to find the right training or resources.

Probably the main reason cross-cultural differences are often swept under the table is the belief that "at the end of the day, we're all the same." Although it's true that there may be universal beliefs and values shared by many team members, to a great degree our respective cultures define how we as individuals make decisions, give and receive information, form relationships, seek affiliation, reach consensus, argue ideas, assess trust, write reports, juggle priorities, and a host of other habits and behaviors that affect collaboration and team communications.

Likewise, it can be dangerous to employ a "one-size-fits-all" approach when team members represent multiple generations. In fact, for the first time in the history of the workplace, organizations need to accommodate the contrasting communication styles of four distinct generations. This chapter offers tips to help virtual leaders navigate through both cultural and generational differences that are most likely to get in the way of effective communication and collaboration.

In this chapter is a global communications tip sheet (Table 7.1) that covers relationship building, written and oral communications, and creating and delivering presentations. These might be especially helpful for those who are new to leading global teams, or those who want to avoid some of the most common cultural differences that have a way of tripping up global teams.

7.1 Galvanize Global Virtual Teams with Clear Operating Principles

The best way to get a new team out of starting gate is to pull everyone into one room for a few days to carve out goals, hammer out differences, develop team norms, and agree on deliverables, schedules, and roles. Investing in this process allows a team to get through the "storming" phase quickly. In today's global organizations, however, most teams don't have the luxury of face-to-face (FTF) bonding time. Team members tend to work across countries, time zones, and organizations. They must rely on remote interactions, both synchronous and asynchronous, to mobilize the team and get everyone moving in the right direction.

Virtual teams that invest time and energy up front in creating explicit operating principles and team norms stand a far better chance of moving forward more quickly than teams that omit this critical step. Here are some tips I co-authored with LeaderGrow, Inc. president Robert Whipple, focusing on a few crucial areas that global teams most need to pay attention to when articulating their operating principles. (See Chapter 4 for more on best practices team norms for virtual teams.)

Table 7.1   Global Communications Tip Sheet

KEYS TO SUCCESS FOR VIRTUAL TEAM LEADERS: THE BASICS

Listen! Watch (if you can)! Observance is the most effective communications skill of all. If you're in doubt about the proper protocol, listen and watch (if you're using videoconferencing or webcams) closely for cues, which may be subtle. If you're confused, it's usually better to ask than to assume you've guessed it right.

Learn the languages (at least a few key phrases): Don't be daunted if your accent isn't impeccable. Your team members will appreciate any attempt to meet them halfway. Start with the basics (hello, thank you, excuse me, good morning/evening) and build from there.

Know your audience beforehand: Take the time to understand important cultural factors that may affect your business or personal relationships. You can collect information many ways, including:

- Talk to those who have been there, or who are natives of that country. Ask them what they see as the greatest differences between your cultures, both personally and professionally.
  Go online and scan the local newspapers or magazines to get a feel for noteworthy current events, cultural values, and possible topics of conversation (avoid politics at all costs).
- Pick up a language book to learn key phrases.
  Read an historical novel of the region or area, which tends to convey more nuances about the culture than a nonfiction book.
- Watch some films made about that country, or by the country's filmmakers, to get a sense for the values, means of expressions, and other important elements that can help you communicate more effectively.
- Search online for articles and guides about the countries of interest. Subscribe to RSS feed, blogs, and other online resources.

Familiarize yourself: Learn the status/rank, level of influence, educational background, relationship to others, and other important characteristics of as many key people as possible. It's always helpful to learn as much as possible in advance, to avoid potential embarrassments later.

Punctuality: Most Northern European countries place great value on being on time, especially for business meetings. (Latin-based cultures are considerably more forgiving if you show up a few minutes late.) Make sure you factor in these cultural conventions before you decide whether to start the meeting on time, or wait for latecomers.

RELATIONSHIP-BUILDING: TIPS AND TECHNIQUES

Introductions and formalities: Because there are as many variables associated with introductions and other formalities as there are countries of the world, take the time to learn the preferred customs. Don't assume that first names will be warmly received, unless you know that to be true about a particular person. In many countries, first names are not used until a good relationship has been established. This is changing with younger generations, and varies by company culture, to some extent.

Make sure you're familiar with the correct (preferred) title. For example, PhDs in most other countries of the world go by "Doctor" (or Herr Doktor) versus Mr., Ms., Miss, or Mrs. (or Herr, Frau, or Fraulein). Similarly, occupational designations, such as "Professor" are often used in place of, or in addition to, Mr., Ms., or Mrs.

(Continued)

Table 7.1 (*Continued*)   Global Communications Tip Sheet

RELATIONSHIP-BUILDING: TIPS AND TECHNIQUES

The art of "small talk": In some cultures, small talk is essential for building a trusting relationship. Some, such as Southern Europeans, Latin Americans, and Asians, find it imperative to get to know their business associates on a level other than business, before a solid foundation may be built. Others, such as Northern Europeans and Americans, often find such chit-chat to be unnecessary and distracting to the business at hand. Know the difference, and be prepared to either indulge in small talk, or skip it altogether. Build in the appropriate amount of time in your virtual team meetings as well as your 1:1 sessions.

Use (or nonuse) of humor: When it doubt, don't! Many Americans are fond of opening meetings with some light humor, as an ice-breaker, or a way to defuse tension. Humor has little place in business settings in many parts of the world. (And even when it does, it's critical to know what kind of humor is best employed, and what topics are best avoided.) If you feel compelled to add a bit of humor to your presentation, check with a local colleague to make absolutely sure it will have the intended effect.

CAN YOU PUT THAT IN WRITING?

Among most Northern European countries and in the United States, business people look for extensive documentation loaded with details, facts, and figures. In other European countries, such as France, meticulous documentation may be considered overkill. Corporate norms also will affect the frequency, content, and details expected for written communication. Here are some general tips:

- Test for possible interpretations: Before you send out that important document, ask someone fluent in the business language of that country to check the meaning against your intended meaning.

- Formal versus informal: This will vary according to both country and corporate culture. Make sure you understand the accepted norms, and tailor your correspondence accordingly.

- Speed of response required: Your response time can set the tone for the relationship to come, especially if your first communications are in written form. Be prompt, be thorough, and be clear in your responses, especially if the recipient is not fluent in your language.

- Watch out for currency denominations, dates, and other confusing terms that differ from region to region:
  - If you're using U.S. dollars, say so; many other countries of the world use dollars, too (such as Australia and Canada). People in most countries would prefer a translation to their own currency, as well.
  - Dates are often written differently than we are used to seeing them. In many countries, October 2, 2013 would be written as 2 October 2013, or abbreviated as 2/10/13. Make sure you know which convention is being applied. It's best to spell out the month in your correspondence, lest you cause unnecessary confusion that could mean major problems later on.

MAKING YOURSELF PERFECTLY CLEAR

Clear speech: Enunciate clearly. Many accents are difficult for nonnative speakers to follow.

Speak slowly (not loudly): Just because your audience doesn't speak fluent English, it doesn't mean they are hearing impaired. Take the time to emphasize key syllables, and leave plenty of pauses so your audience can do the mental translation back into their own language. Don't be afraid to check in with them from time to time as to your pace.

Table 7.1 (*Continued*)   Global Communications Tip Sheet

MAKING YOURSELF PERFECTLY CLEAR

Simplicity: The KISS principle applies in most other countries, where elegance and a clear sense of the big picture is more valued than a mass of details (except in some countries such as Germany, where details are paramount). Avoid compound sentences or elaborate statements. Parcel out your ideas in bite-sized pieces that can be easily digested and assimilated.

Active verbs: Use as many as possible to convey your meaning. Passive sentences tend to confuse and dilute meaning. Avoid using the word "get," which means nothing and everything in "Overseas English."

Paraphrase: This is a useful technique to check if the intended meaning has been properly conveyed, without embarrassing yourself or anyone on your team. When we communicate virtually, we have to be vigilant about testing for understanding or shared meaning. Asking, "Anyone have questions?" or "Did everyone get what Juan just said?" will likely result in silence. Instead, take the initiative to paraphrase important ideas to give everyone an opportunity to hear the idea twice, in different ways. Example: "I think Juan just made a key point. I want to repeat it, using different words, to make sure I understand it correctly. Juan, please let me know if I have missed anything."

Avoid acronyms and idioms: We tend to pepper our speech with idioms and allegories that are well understood in our own countries, but may be utterly confusing to someone from another country. In the United States, sports allegories have widely infiltrated our everyday speech, often unconsciously. Consider the phrases so commonly used in the United States, such as "out in left field," "the whole nine yards," and "full court press." Imagine the confusion of an audience whose major sports do not include baseball, basketball, or American football! Avoid local idioms and trendy business buzzwords whenever possible. What's a widely embraced term in one country may only cause confusion in another. It's best to double-check both spoken and written words with a resident of the targeted country.

DESIGN OF PRESENTATIONS

Bite-sized pieces with frequent breaks: Whether you're using an interpreter, or going it alone, allow plenty of extra time to present to nonnative-speaking audiences. You'll be speaking more slowly than usual, paraphrasing to check for meanings, and giving your audience time to do the mental translations. A good rule of thumb is to plan to spend about 50–100% more time than you would to an English-speaking audience. Allow for frequent breaks if you're presenting for more than an hour at a time. Translation is mentally exhausting; your audience will need time to refresh their minds.

Visual aids: Be sure that any visual aids reflect a sensitivity to the local culture. One U.S. client I worked with used a basketball hoop as his central metaphor throughout an entire slide presentation that he intended to use around the world. He got as far as Canada, when he realized that basketball analogies had little relevance outside the United States. Another client wanted to show the Statue of Liberty as a "universal symbol of freedom," until it dawned on him that for people outside the United States, the statue may hold no such symbolism.

Although slides shouldn't be cluttered with a lot of text, they should contain some brief text to amplify your points. Non-English speaking audiences will appreciate the ability to verify that what they think they've heard you say is in fact what you meant to say. Many people find it easier to follow the written word than the spoken word. Provide them with both, augmented by graphics that convey your primary messages with simplicity and clarity.

(Continued)

Table 7.1 (*Continued*)   Global Communications Tip Sheet

DESIGN OF PRESENTATIONS

Level of interaction: You may find yourself being interrupted by a steady flow of challenging questions or philosophical disagreements in France, whereas in Korea, the most interaction you might receive from your audience is a polite question. When you're planning your team meetings, think about how you'd like your audience to interact, and what you can reasonably expect, given the culture. Many presenters like to lace their presentations with questions—or challenges—to the audience. Be prepared to forego your usual style, if the norms of the local culture don't call for a high level of interaction between presenters and audience members.

To sum it all up: Most audiences will look for your major points to be restated at the end of your presentation. Don't disappoint them; make sure you clearly convey your key points at the end, even if it is the third or fourth time you have made these same points.

- **Selecting the right people:** In the formation of an intercultural remote team, the selection of individuals is critical, assuming you have a say. Establish a list of needed characteristics for team members, and screen all proposed members with care. For example, ideal global virtual team members have an understanding of and tolerance for cultural differences that can affect work patterns. They are also flexible with respect to work hours and feel comfortable using multiple forms of communications. Effective virtual team members have a high tolerance for ambiguity and can work easily with little direction.

- **Setting realistic expectations:** Think about the areas in which unrealistic expectations are most likely to trip up the team, such as deliverables, schedules, resources, and support required. As a team, clarify mutual expectations and validate whether they are realistic. Create a means by which members can reset expectations when needed. Remember that some cultures are more conservative when setting expectations and need to consider the big picture first. Other cultures tend to be overly ambitious in their promises and need frequent reality checks by others to ensure they're really able to deliver.

- **Establishing a "safe" environment:** The online environment offers the opportunity for people to raise issues, ask questions, offer opinions, or contribute new ideas without inhibition. By establishing a culture where everyone feels encouraged to contribute openly, the team leader can tap into a rich diversity of ideas. Agree on ways people can ask for help without embarrassment, and determine how concerns can be surfaced without fear of retribution. Know how different cultures are likely to assess the meaning of "safety." Consider anonymous forms of participation to lower inhibitions, especially when topics are regarded as contentious or difficult.

- **Handling conflict:** Consider what type of conflicts are likely to arise, particularly those that may be most difficult to handle remotely. For example, what if one member consistently fails to fulfill commitments? Or what if vital information is not shared openly with all? Agree how certain conflicts will be handled, who will be involved, and what communication method will be used. Make sure to reflect the cultural makeup of the team when thinking through your choices. Some cultures are more comfortable being direct and assertive whereas others place greater value over group harmony than speed of execution.

- **Varying time zones:** When working in a synchronous, or real-time, mode (instant message, web meeting, telephone, video conference), some remote team members are forced to work at awkward times. Agree as a team when same-time meetings are necessary, and consider rotating the times to share the burden of working during normal sleep time. Consider which work can be done asynchronously (e.g., via e-mail or a shared workplace) to allow all team members to work at the most convenient times.

- **Making decisions:** Decision rights need to be discussed so there are no surprises. Once the lines of authority have been established, all team members need to remain compliant or people may start playing political games, which will quickly undermine trust. Intercultural teams must accommodate the various negotiating styles of different members. For example, many western cultures call for "first, best offer" tactics, whereas some eastern cultures expect some dickering. Keep in mind that many virtual teams have no obvious leader, which makes agreement regarding decision rights more challenging and more critical.

- **Reporting relationships:** With virtual teams, the reporting structure is rarely static. Team leadership can morph as the nature of tasks changes during the life of the team. All team members need to feel that they are contributing with maximum creativity and energy. Learn how different cultures are likely to regard the relative importance of hierarchy, authority, and reporting relationships. For some, it's sacrosanct, and for others, it's almost completely irrelevant. (Of course, certain functions and organizations have their own values assigned to notions of hierarchy and power, which must also be considered.)

- **Project scheduling:** When mapping out a project plan, keep local customs in mind. For example, include all national and religious holidays. Consider planned vacations as well. In some countries, especially in Europe, team members will be taking extended vacations in the summer. Some cultures accept working on weekends or during vacation times; however, many others do not. Make sure that the entire team works from a shared calendar, which should be posted in a place to which all members have ready access. Encourage team members to block out time well in advance on their electronic calendars to make scheduling across borders and time zones easier.

- **Tracking progress:** Many team activities will span a significant time period. To give the team a greater sense of progress needed to maintain momentum, try tracking progress against subgoals. This way, the team can modify tactics if necessary and can celebrate many successes—small and big—along the way. Using some type of visual barometer, such as a dashboard or thermometer, can be especially helpful for a global team when not all members have the same proficiency in the shared language. Make sure everyone understands the criteria for reporting progress. For example, if a dashboard is used, agree on what "green" really means.

Global teams need to learn how to operate successfully in a virtual world, navigating through cultural differences which, if ignored or dismissed, can easily thwart progress. Organizations that know how to nurture and support global virtual teams have a competitive advantage over organizations that treat all teams and their members the same.

7.2 Real Cultural Assimilation Takes Patience, Time, and Willingness to Adapt

Many virtual team leaders have been fortunate enough to have had at least one opportunity in their professional careers (or college days) to live and work in another country. For many organizations, in fact, the rotation of managers through functions, regions, and roles is a vital element of their talent management strategy.

When I took a post in Hong Kong in the early 1990s, I had relatively few resources at my disposal to learn more about the business and social cultures of the countries in which I would be working throughout Southeast Asia. I interviewed a couple of people, read the relatively few intercultural business books then available, and went off to my assignment on little more than a wing and a prayer.

Today's leaders have many more means by which they can learn about other cultures before they become immersed. Cross-cultural books and videos are ubiquitous, as are training programs (both online and in-person) and reference guides. In addition, those who work as part of a global team can tap teammates for a quick primer on aspects of their culture that are intriguing or confusing. Still, successful cultural assimilation takes extraordinary patience, superb listening skills, and exceptional powers of observation. Here are tips I wrote with my colleague Clint Cuny of Export Trading Group USA, to help leaders and members of cross-cultural teams to develop rich relationships that lead to successful collaboration, more quickly.

- **Take the first best guess.** It's true that stereotypes can unfairly bias us toward a particular group of people. Our predispositions are a collection of our personal and professional life experience. But it's also true that making some generalizations about a new culture can help us avoid some costly missteps early on. For example, if we know that in a particular country, employees typically defer to their leaders as experts, we may not expect a brainstorming session to yield many responses if the leader is present. Or if we learn that some cultures are reluctant to make decisions without first cultivating a personal relationship, we can calibrate our conversations accordingly. Without some well-informed generalizations about a culture, we typically revert to approaches that are comfortable for us, a tactic that's sure to backfire.

- **Understand how underlying values drive behavior.** Think of culture as an iceberg where observable behavior is the 10% you can see and 90% is hidden. It's one thing to observe a behavior, say, that workers are generally respectful about taking direction and don't make it a habit to check in until their work is due. Unless you understand the cultural factors at play, you may unintentionally offend your colleagues if you request periodic progress reports because you can't stand to be left in the dark. Discovering cultural values takes time. Try reading books or watching films to augment what you learn from your colleagues, who usually do not take an anthropologist's view when explaining their own behaviors.

- **Listen and observe before acting.** Pay careful attention to interactions among people and groups and seek understanding of what is really happening. Who's making eye contact with whom (assuming you have access to visuals)? Is there a usual order in which people tend to speak? How much direction do people seem to need or want? Are meetings considered open forums in which people can ask questions and debate, or are they led by senior leaders with little room for discussion? Take copious notes, including questions you have as you observe interactions unfolding. Find a time and place in which you can sit down with someone you know and trust to state your observations and ask open questions about what's intriguing or puzzling to you. Try scheduling regular "cultural coaching" sessions with one or more colleagues, especially early on.

- **Pace yourself.** Americans, especially those of us from the northeast United States, tend to operate at a pace that others find unnecessarily frenetic. People who work at a more deliberate pace can drive us crazy. Discover how the other culture regards time from a number of aspects, such as punctuality, levels of responsiveness, decision-making processes, and planning horizons. Determine how you'll have to adjust your own style or create strategies for adjusting to theirs. (For example, if you have a low threshold for being kept waiting, try bringing something to work on if you suspect your colleague may be tardy. If you tend toward snap decisions, postpone announcing decisions, lest others think your decision may not be well thought out.)

- **Adjust to different work habits.** In some environments, people work heads-down for 12-hour days, up to six days a week. In others, people sashay in a tad too late for your taste, take a long leisurely lunch, and then leave an hour or two later than you may be used to. We have seen some managers try to impose their own "work ethic" on people of other cultures. To avoid mutual frustration and distrust, be prepared to adjust your expectations about the hours you can realistically ask others to put in, based on local cultures and labor laws. If you're used to persuading team members to work weekends or during vacation time to meet a critical deadline, don't assume that will work in this case. Plan projects to safeguard personal time.

- **Practice patience.** Many Americans assume, implicitly or explicitly, that our penchant for speed is necessarily valued by other cultures. We tend to move "full speed ahead" in seeking input and consensus from all, gaining rapid agreement, and asking for direct feedback. We also move to quickly knock down any barriers that get in the way of speedy progress. In reality, however, many cultures take a more deliberate approach and feel uncomfortable being rushed to premature conclusions. Take your cues from your colleagues, and err on the side of moving more slowly than you may feel necessary to gain trust and win allies.

- **Relationships matter.** For some cultures, trusting relationships can be a prerequisite for doing business and matter more than achieving a particular business outcome. Many Americans value efficiency and speed in achieving "results" so much that they often overlook the importance of personal relationships to others. (After all, it takes time to grow a relationship, which is simply not very efficient!) Rushing to conclude an agreement or make a speedy decision inevitably causes the opposite to happen, especially in cultures where building a trusting relationship with the other person is a necessary first step.

Even if you take the time to learn everything you can about another culture before you immerse yourself, you'll need large reservoirs of patience and understanding to eventually succeed. (And no, success will not come overnight.) Add a day or two to the time you'd normally expect a report. Don't complain if people don't join conference calls (in some countries, such as in Africa, conference calls can be inordinately expensive and facilities don't exist in many locations). Know that not everyone will keep to the agreed-upon schedule.

There is a saying among Americans doing business in Africa: "TIA," meaning "This is Africa." Change that letter "A" to whatever culture or country in which you are working and remind yourself that you need to work hard to adapt to their culture, and not the other way around. In the end, most people are decent, are eager to do the right thing, and want to be respected. Our interactions should reflect these principles. So watch, observe, don't anticipate, and leave your predispositions at the proverbial door.

7.3 Surfacing and Addressing the Cultural Differences That Most Affect Virtual Teams

Some aspects of teamwork tend to suffer more due to cultural differences that are ignored or dismissed. Here are some aspects which, if successfully addressed, can catapult a global virtual team forward surprisingly fast, once they get through the tough but necessary conversations.

- **Decision making:** Some cultures (notably Americans) are known to value speed above all when it comes to making decisions, even if it means they're made with incomplete information and insufficient buy-in. The unfortunate result is that decisions often must be revisited and recast, leading to costly rework. Other cultures—Japan, for example—tend to be more holistic in their thinking, requiring considerable time to assess the rationale and impact of decisions, methodically seeking buy-in from a variety of stakeholders. Decisions may take longer, but implementation comes faster. When you're part of a virtual team, it's important to articulate operating principles about how this team will make decisions, including timing, criteria, process, approvers, input required, communication of decisions, and so on. Different types of decisions might require different principles. Be prepared to engage in some spirited debates as you get to common ground.

- **Information sharing:** Some cultures share and request information freely, up, down, and across the organization, without regard to hierarchy. If they need information to get their jobs done, why stand on ceremony if it means an avoidable delay? Generally speaking, in the United States, open sharing is the norm. Some other cultures tend to parcel out information on a need-to-know basis. Information is compartmentalized and funneled along functional or organizational lines. Because same-time conversations are rare for most virtual teams, members need explicit agreement about how information will be shared. For example, what kind of information will be posted on the team's SharePoint portal versus sent via e-mail? Who has access? To create this kind of "information architecture," team members must spell out for each other what information they most need, at what point, to get their work done.
- **Level of participation:** Formal cultures that place value on hierarchy and seniority may not be as willing to assess an idea in front of others, especially if a senior manager is present. Team members may tend to wait until spoken to, and even then, may not offer any contradictory or critical views. Other cultures may enjoy a lively debate and in fact relish the idea of proffering opinions to anyone who will listen, without fear of any negative repercussions. Team leaders need to be sensitive to these dynamics and carefully plan their meetings to accommodate differences. For example, if people of different seniority levels are on a call, make sure that junior people have a comfortable way to participate, such as by providing a web meeting tool that allows for anonymity. Also consider how you'll coax quiet participants to speak. Some respond well to being called on, whereas others resent the attention. Find ways to encourage lively participation from everyone, even if it means providing different tools, at different times.
- **Motivation and rewards:** Some cultures don't seek out or expect recognition or rewards, and are inherently gratified by simply doing their jobs well. Other cultures, however, often expect some type of reward, monetary or otherwise, for meeting their goals. In addition, some cultures tend to value individual recognition, whereas for other cultures, it's the team

effort that people like to see rewarded. When considering how best to motivate, reward, or recognize a cross-cultural team (or its individual members), realize that you need not always have a one-size-fits-all kind of reward or recognition that works equally well for everyone. Also pay attention to local laws and norms, especially when deciding upon any type of monetary reward.

- **Punctuality and deadlines:** Everyone's schedule is jam-packed, with people on back-to-back calls throughout the day (nights, too!). That's why it's especially important to establish a team culture that values punctuality for virtual meetings, even when a member's cultural predisposition is to show up "whenever." Set ground rules about punctuality and stick to them (e.g., "Latecomers catch up on what they missed afterwards.") Likewise, some cultures want precise deadlines and hold these sacrosanct, and others see deadlines as a goal that can be flexed depending on the circumstances. "Status reports due next week" may mean Monday at 9 a.m. CET to some and Friday at "end of day" EDT to others. Make sure everyone agrees when important deliverables are due, making clear the impact of slipped deadlines. If some deadlines matter more than others, say so.

- **Policies and procedures:** Some cultures like to tackle one task at a time, completing one before moving on to the next. Logical, sequential, well-defined processes are necessary conditions of work. Other cultures regard a frenetic work environment and frequent interruptions as vital and even welcome. Relationships come before processes, and distractions offer unplanned opportunities to learn. For a virtual team, acknowledging these differences and deciding how to address them as a team is crucial. For example, some may turn off e-mail a few hours a day to maintain focus; others like to send IMs whenever the mood strikes, and get frustrated when they don't get an instant reply.

- **Work/life balance:** Extended vacation times for some team members may be resented by those who have to pick up the slack, especially during crunch times. In general, Americans tend to have far less vacation time than their European counterparts, and are more likely to work through weekends or holidays if that's what the project takes. Before ill will can fester due to this perceived inequality, discuss principles and values regarding work–life balance, including what's acceptable to ask and what's out of bounds. Discuss how the negative effects of prolonged absences can be mitigated. Make sure that everyone has a world calendar so everyone can plan around local holidays like the 4th of July in the United States, New Year's Week in China, or Bastille Day in France.

When working as a virtual team, cultural barriers tend to get magnified. We often revert to viewing others through our own cultural lens and often see other cultures as a "not-quite-right" version of our own. Be actively curious about how cultural differences are affecting your team in both positive and negative ways. Learning about other cultures is only half the equation. Find ways to learn how your culture is perceived by other cultures. When in doubt, enlist a "buddy" from another culture with whom you can check assumptions, get feedback, and ask for advice as you find your way along.

7.4 Communicating across Cultures: Designing for International Transportability

As a virtual leader, you may sometimes be charged with creating global communications strategies beyond those of your team. Let's say your organization is poised to launch a new program that will rock the world of almost everyone across the organization, in all regions of the world. You've spent weeks with marketing, HR, and legal to hammer out a set of crisp consistent messages. At last, you have sent the content to your team members around the world, leaving you time for last-minute tuning and tweaking.

Just as you're about to breathe a sigh of relief—prematurely, it turns out—you receive a stream of heated e-mails asking, essentially, "What on earth were you thinking?!" You're shocked and confused. After all, you worked weeks to create core messages that could be translated from English without losing much meaning in the process. And what about that multichannel communications matrix that took you days to put together, designed to accommodate different cultural preferences?

Here are seven key steps that any virtual team leader needs to take in creating and implementing a global communications plan designed to resonate with those most affected by the change.

- **So, what's the big deal?** The magnitude of any given change will vary by the segment of the employee population, such as job type or role, as well as by location. It's dangerous (and costly) to make sweeping generalizations by polling only the people closest to you. In fact, you need to cast a very wide net at the outset to determine how people in different countries or regions are likely to be affected by the change. Set up time to speak with country contacts, either 1:1 or in small groups. These contacts can act as your "change translators" who can help you to gauge the anticipated level of resistance, or receptivity, this change will engender in their locations. Consider also setting up a virtual conference area where your contacts can contribute to a larger "idea bank" at their convenience. Publish and discuss your results with your project team, as well as with your contacts who have contributed their ideas.

- **Create a context for change.** As you begin creating your communications plan, find out what other changes lay ahead. What is the nature and scope of other changes? What's the likely disposition of affected stakeholders? Timing? To what extent will different project teams have to compete for mindshare? Is there an opportunity for "co-marketing" multiple projects at one time? If you're lucky, your organization might have an "air traffic controller of change" who is aware of all projects rolling out, potential intersections, possible conflicts, and opportunities to streamline. In addition, you'll need to find out what other activities might affect your project at a local level, such as a planned shutdown, a wave of downsizing, or a recent merger.

- **Maintain a global network of "go-to" change consultants.** These people may play different roles from country to country or region to region; for example, they may be members of your extended project team, local communications professionals, HR consultants, or interested stakeholders. Whatever the title, these are the people you will rely on to validate messages, solicit communications advice, pilot communications samples, and generally provide guidance to the project team every step of the way. These may or may not be the same people as the "change consultants." Plan to meet network members frequently, at least weekly in the early planning phases and during rollout.

- **Consider the relative importance of global messages.** Many organizations insist on "globally consistent" messages and in so doing, may generate content that's pretty much useless in other locations. Sure, a company has to make sure that its branding is consistent around the world. But for change that affects an organization's own employees and managers, country and regional representatives need latitude when it comes to interpreting the implications. In addition, the people communicating the changes need to feel comfortable adding their own spin, adjusting for their personal style of communicating. When precise language is required for legal reasons, stipulate that clearly. Otherwise, messages will be far more credible when they can be refined for local influencers and their audiences.

- **Avoid getting lost in the translation.** Assuming you already know which countries require language translation (and who's paying for it), make sure you allocate sufficient time for your local project team contacts to review the translated content first. After all, only those who really grasp the changes and related implications can make sure that the translation is accurate, clear, and preserves the intended meaning. Also make sure you know which groups need content in the local language. Not all employees within a given country need a local translation. As a rule, for U.S.-based corporations doing business in other countries, the more senior the country team, the less likely they are to need content in the local language.

- **Respect the power of local gatekeepers.** The gatekeepers who will be transmitting the messages for local audiences need to be selected with great care. First, they must be regarded as credible effective communicators in their own right. To maintain credibility, they need to have an in-depth knowledge of the project, because they'll need to field questions from local audiences in real time. (Simply providing a set of FAQs to someone who has just a cursory knowledge of the project won't cut it.) In addition to understanding the project inside out, they must be sensitive to the perceptions, fears, and concerns employees are likely to have regarding the change ahead. Poll these local gatekeepers well ahead of time to find out how the project team can best equip them to be successful communicators of change in their areas.

- **Create a flexible communications menu.** At a project level, you may have a core set of communications pieces lined up, such as e-mail templates, FAQs, PowerPoints (PPTs), articles, project blogs, and scripts, all modifiable by local contacts. Interview your local change translators or gatekeepers (who may or may not be the same people) as to what communication devices are likely to have the greatest impact, given this change. In some facilities, for example, tent cards and posters are more effective than e-mail or blogs. In parts of Europe, miniposters in bathroom stalls are common to grab people's attention. A mousepad or mug might be popular in some places, whereas they may be seen as an expensive turnoff in others. Weave a global "tapestry" of communications offerings, indicating the best use of each, and let local contacts assemble the best combination of elements as they see fit.

Invest the time in building trusting relationships with your local contacts. Find people who can act as reliable change translators and communications advisors. Earn credibility by seeking out and incorporating their ideas into your overall project planning. Take time to familiarize yourself with the "senders," or gatekeepers, to find out what makes them tick. The bottom line is to be prepared to spend time on the front end of the project to create and cultivate ongoing relationships with trusted local contacts. Otherwise, you'll need a lot more time on the back end to deal with the likely resistance, confusion, frustration, and costly delays at the time of launch, and perhaps long after.

7.5 Listening and Learning across the Generations: Strategic Communications Planning for Better Collaboration

Regardless of their ages, many managers fail to take generational preferences and styles into account when mobilizing and motivating their teams. Instead, they develop team norms and operating principles that may run counter to what individual members might need or value. Although the different generations go by a variety of names, here we refer to them as Traditionalists (born 1927 to 1945), Boomers (born between 1946 and 1964), Gen Xers (born between 1965 and 1980), and Gen Yers (born after 1981).

For example, a Boomer manager may insist that all people work from a central office during typical working hours. However, many Gen Yers are most productive at 10 p.m., working from the comfort of home. Some Gen Xers, on the other hand, may need an afternoon off for family obligations, coming back online later that evening. Instituting a rigid policy about work hours or locations may leave some team members feeling alienated, excluded, and ultimately, not very productive.

Here are some tips co-authored with my colleague Sheryl Lindsell-Roberts of Sheryl Lindsell-Roberts & Associates to help connect people from different generations through more targeted communications. (Please note that although each person deserves to be treated as a unique individual, making some "best-guesses" about communication styles and preferences is an important first step toward creating a team communications plan that works for most.)

- **Rethink "normal" work hours.** Apart from some government offices and banks, the 9-to-5 business day has given way to more flexible work times and locations, with people working at all hours from multiple locations. For a team that works virtually, it's much harder to find an agreed-upon window for group meetings, whether face to face, phone, web conference, or videoconference. A Boomer manager may feel more comfortable when all team members convene FTF for the weekly 8 a.m. status meeting. But consider a Gen Xer who is caring for a family, and needs to battle traffic for 90 minutes to get there. Or the Gen Yer who insists he is most productive from 11 a.m. to 11 p.m. Many Traditionalists easing their way into retirement are also demanding more flexible work arrangements. Managers must consider the comfort level and preferences of all participants when deciding which team meetings really need to take place FTF and which can be done via conference call or web conference.

- **Sharing vital information.** When time is of the essence and you need to get critical information to team members, what's the best choice? It depends on a host of factors, including the likely preferences and habits of members representing different generations. Older generations tend to rely on e-mail, phone, or FTF as the default, whereas many younger members may look to instant messaging, blogs, wikis, or texting as their primary means of giving and getting important information. Consider multiple channels for information-sharing, especially if you have people with strong preferences for different communication methods. At the same time, make sure you have an agreed-upon method for sharing urgent information, such as news likely to affect the work of the team or missed deliverables that will trip up others. Keep in mind that younger generations tend to be natural and eager collaborators, and often do so from a distance as a matter of routine.

- **There's no place like "home."** Create a team portal that's easy, quick, and intuitive for members of all generations to use. Younger generations expect and demand highly efficient websites where needed information takes just one or two key clicks to find exactly what they need. Otherwise they may tune out quickly. Older generations may require a bit of prodding to regard the team portal as the place to go to share and view the latest and greatest information. If people are slow to gravitate to your team portal, try pushing out e-mails that contain a sentence or two about what content can be found on the team site, and refrain from including the actual information in e-mails so they have more incentive to visit the portal. Constantly seek feedback from team members representing all generations as to how the team space can be made even more useful.

- **Instant gratification versus patience as a virtue.** Younger workers typically expect responses and information right now, as evidenced by the surging use of social apps of all kinds, and the proliferation of text messaging and IM over phone and e-mail. Waiting a day or two to receive a return e-mail or voicemail is a nonstarter. Older workers tend to expect a reply to take a little more time, and likewise may be slower to respond themselves, especially if they have to wade through a jammed inbox to reply. Create agreed-upon norms for responsiveness to certain types of inquiries or issues, and then determine how best to use specific tools to get the job done. If one person insists on an IM or a blog update and another prefers an e-mail, work together to agree on the best ways to meet as many needs as possible without extraordinary effort from any one.

- **Wide open versus buttoned up.** Older workers tend to prize consistency, predictability, accuracy, good grammar, and thoroughness in communications. Even the most creative ideas may be dismissed if such ideas crop up randomly, without context and without a way to prioritize them. After all, if the ideas don't lead to something tangible, you've just lost a lot of time! Younger workers, on the other hand, are adept at brainstorming and collaborating with people who have shared interests, including total strangers, for the sheer joy of creating something new and fresh. Social networks enable this type of spontaneous collaboration that may lead to great new ideas that may do nothing more than satisfy intellectual curiosity. Openness and creativity are especially valued by younger generations versus playing by prescribed rules of engagement, which is something their parents may be more prone to do.

- **Ramping up and ramping down.** People from different generations have a lot to teach each other, if we create the right opportunities for knowledge transfer. Many younger people coming on board bring rich new perspectives, a keen appreciation of how best to apply the right technology tools, and a passion to learn. They embrace challenges with gusto and are devoid of the "this-is-the-way-we've-always-done-it" mindset. Older people, many of whom might be nearing retirement, have accumulated wisdom about the business, industry, and organization, and know what it takes to operate successfully within the enterprise. Two-way mentoring programs, pairing a younger employee with a more senior counterpart, afford the opportunity for both to learn from each other. The result is that new people position themselves for success more quickly and older workers can leave behind valuable knowledge skills and knowledge for the next generation.

The bottom line is that organizations need to examine the most significant generational differences and determine how best to anticipate and address the implications within each work team. The outcome is high-performing teams that consciously take advantage of generational differences instead of ignoring or dismissing them.

7.6 Open Communication and Mutual Respect: Keys to Intergenerational Harmony

With multiple generations working side by side for several years now, we have learned a lot about the differences most likely to affect the ability of multigenerational teams to collaborate successfully. Some organizations have taken this advice to heart and consciously work to reflect these differences when it comes to selecting and cultivating teams. Others have dismissed the advice as irrelevant, unimportant, or simply too overwhelming to do much about.

Most of what's been written has come from those of us who are considerably older (and more experienced) than our Gen X and Gen Y counterparts. Here my colleague Sheryl Lindsell-Roberts and I sought the perspectives of some of our Gen X and Gen Y colleagues. After all, for all of the wisdom we older generations think we have to offer, the Gen X and Y folks of the world have a lot to teach us, too. Here are observations and advice culled from interviews with our younger counterparts:

- **Take the time to teach us.** We know that you have wisdom we can benefit from, but sometimes we don't know the best way to get at it. We know, for example, that a lot of thought must go into making some of the tough and complex decisions you need to make, but unless we understand the logic and rationale, we can't learn from you. Plus, we may not be happy with some of the decisions you make if we can't understand why you made them. Ask us for our input. Involve us in making decisions when it's appropriate. Find time to spell out for us the reasons you say and do the things you do so we can capitalize on your knowledge. And who knows? We may have ideas of our own that you can use as well.
- **Give us the opportunity to teach you.** Even though we may be relatively new to the business world, we have interesting

perspectives and fresh ideas to offer. Just ask us. And if you don't know how to use instant messaging, or if you feel uncomfortable texting, or if you have reservations about using wikis, blogs, or social networks, we can show you how. We know that some of you are much more comfortable talking face to face or using the phone, and we respect your choices. But we ask you to open your minds to trying new avenues so we can all feel more confident and comfortable communicating.

- **Let's appreciate each other's communication styles and preferences.** You have a way of communicating that's typically more formal than ours, and we know that this is a quality we have not mastered. We, on the other hand, tend to favor quantity over quality, given how many devices we constantly use to send and receive messages. We are more casual and familiar, and we like to use cryptic abbreviations, emoticons, or pop culture expressions as a way to build relationships. Don't assume this means we're immature or disrespectful. This is simply how we build bridges. If you ask us to be more thorough or more clear, we may complain that you're slowing us down, although we realize that this could be a good thing at times. Let's have an open discussion about what styles and preferences work best and under what conditions. If our business requires a certain set of norms we can all agree to, so be it. But if it's simply a matter of communication preferences, let's respect the fact that each of us has styles and methods with which we're comfortable.

- **Don't make distinctions between cultures and classes.** We may not have seen as much of the world as you have, but our generation doesn't seem to notice or care as much about the differences between this culture or demographic and that one. We're used to operating in a global arena, and we tend to create our own communities of interest that span countries and cultures. So when you talk about the communication styles or attitudes of people from this country or that region, we don't often experience those differences in our own communications with people from those areas.

- **Tell us what information you need, how you want to receive it, and why.** We are adept at collecting information from a hundred different sources and putting it all together as fast as we can. You may complain that our reports seem half-baked or superficial, and we may feel that your information requirements are burdensome and pedantic. If there are good reasons for your requests, such as regulatory requirements or the need to adhere to standard financial reporting procedures, let us know. But if these requirements have simply been passed on as a result of some reporting structure that made sense 20 years ago, let's work together to create requirements that make sense for the business today.

- **Realize that multitasking is not all bad.** In fact, using a bunch of different communications devices at any given time is a way of life for us. We are not necessarily being disrespectful when we don't focus 100% on a given conversation (although we can see how you might think we are). It's just that we're hardwired to receive and relay information constantly, and it's hard for us to suddenly stop. Why not find ways to put our penchant for multitasking to good use, such as incorporating a variety of communication tools into our real-time meetings? If our multitasking is getting in the way, then we ask you to find better ways to engage us by encouraging our ideas and active participation.

It's important for all of us to remember that all the "rules" that once governed the workforce no longer exist. Today there are no rules. The companies that thrive (not merely survive) are the ones that respect and harness the potential of today's rich and diverse multigenerational workforce. This increases the bottom line, serves customers better, and makes for a cohesive working environment.

7.7 Summary

One of the most rewarding aspects of being part of a global virtual team is the rich diversity of perceptions and experiences that different cultures and generations bring to bear. Thanks to an ever-flattening world, we have opportunities to collaborate and learn from people we may never meet face to face. The challenge for the virtual leader is to create a team environment where cultural and demographic differences are openly discussed and differences are bridged, thereby creating a team culture that blends and transcends multiple cultures and generations.

8

TROUBLESHOOTING TIPS
FOR VIRTUAL TEAMS

Let's face it. All teams run into trouble now and then. Even those that are humming along beautifully for months at a time can suddenly veer off the rails without so much as a warning (or so you thought at the time). Sometimes the signs are obvious, and team leaders have time to make effective interventions to head off small issues before they grow to be debilitating problems. Other times, the problems are more insidious and nuanced, making them harder to spot.

For virtual teams, discovering dysfunctions both large and small can be very tricky, even for the most astute listener. In the absence of visual cues, team leaders can't easily tell whether those prolonged silences from a particular team member are a sign of ennui, frustration, or simply maniacal multitasking. If a couple of team members repeatedly fall short of commitments, it's harder to figure out if there's been an innocent misunderstanding, or whether there are more serious issues at play that need to be discussed. Discovering the existence of issues from afar is one thing. Successfully addressing them when working remotely is quite another. In this chapter, we offer tips for troubleshooting virtual teams when things go awry, either a little or a lot.

8.1 Six Management Practices That Don't Cut It in a Virtual World

Managers who have demonstrated impressive leadership skills working with co-located teams often mistakenly assume that as seasoned management professionals, they'll automatically excel working in the virtual world. In fact, it's the managers who presume instant competence when moving to a virtual role who struggle the most. That's because they have not felt a burning need to seek out new skills, erroneously assuming that leading any kind of team pretty much requires more or less the same skills and time-honored approaches.

If you have been wondering why your own proven management approaches don't seem to be hitting the mark with your virtual team, here's why, in a virtual world, some "traditional" management tenets backfire, along with alternatives that will yield better results.

- **Don't trust what you can't see.** Leaders who micromanage mistakenly believe that the more one checks up on people, the faster they'll produce results. Not so in a virtual world, where this kind of pestering may force team members to pull back or pull out, many times with no one really noticing. Far better to pick up the phone or send an e-mail or IM to ask how things are going, using a friendly and supportive tone. However, do this sparingly, as people may start to see this show of alleged concern and support as micromanagement in disguise.

- **"Because I said so."** A command and control style rarely works outside the military in any work environment, but in a virtual world, it's a nonstarter. Some managers imagine that people will be motivated to perform high-quality work just because their manager declares that they must. Hardly. In a virtual world, where it's so much more challenging to ensure that a team is aligned to work toward shared goals, it's critical that everyone buys into the overall goals, business case, and the context for both their team and their individual contributions. Without such an explicit agreement, team members may work at cross-purposes, if they work at all; and when the lack of alignment is finally spotted, it can be too late to pull them back in.

- **Keep important information close to the vest.** Think that holding back vital information will make you more respected and powerful? Think again. One of the most critical roles of a virtual team leader is to ensure that people have a way to share and access the content and knowledge they need to do their work. In the absence of informal ways of exchanging ideas and knowledge, virtual teams require multiple channels and methods for creating, sharing, accessing, and building on information. Creating a team information architecture is ultimately the responsibility of the team leader, with input from the team. When in doubt, err on the side of enabling members to access more information rather than less.

- **Zero tolerance for mistakes.** Making examples of team members who slip up, miss a deadline, or otherwise disappoint can create a culture of fear and distrust. Creating an environment where people feel pressured to appear perfect may lead team members to hide real problems, fail to surface critical issues, or pretend to be meeting goals and achieving deliverables when they're really not. In a virtual world, it's harder to discover what's going unsaid, because team members have few means by which to share what's really going on. To foster an environment where honest conversation can flourish, team leaders must find supportive ways to encourage people to acknowledge shortcomings without fear of retribution.

- **Do as I say, not as I do.** If you're a typical manager, you may arrive at your meetings late, expecting to be caught up on what you missed. You might then sidetrack the agenda with unplanned topics as you silently multitask, hoping no one notices. Yet, when a team member demonstrates this same behavior, you may show little tolerance. Leaders of virtual teams must model the kind of behavior that enables real collaboration. Otherwise, your team members will start showing up late, checking out early, and participating halfway, simply because it's so easy to do when no one can see you, and when the boss has set the precedent. Set the standard for the behavior you expect from others. If you must arrive late, apologize in advance and catch yourself up on the proceedings through active listening.

- **Light a fire under people and they'll do great work.** Many managers use magical thinking when making unrealistic demands of their teams; for example, "If I set a really ambitious deadline, people will somehow find more hours in a day to get the work done on time." The result, most often, is damaged morale, frustration, erosion of credibility for the team leader, and ultimately, the inability of the team to deliver timely work. When deliverables are slipping, sounding alarms can be difficult and scary for virtual teams, especially when the leader refuses to back off from a preposterous deadline. Far better to ask the team what they can deliver by when and what they need from you and each other. Agree on ways to track progress and identify gaps so all can quickly discover impediments that require quick rectification.

8.2 Avoiding the Unintended Consequences of Micromanagement

How do you know that your team members are working as efficiently and effectively as they could be? How can you be sure they're really on target with their committed deliverables? One of the toughest challenges virtual team leaders face is knowing when to check in or otherwise intervene, and when to pull back to see what happens.

Without the ability to informally check in, virtual team leaders often must rely on formal communication channels to be apprised of deliverables, such as weekly 1:1 check-ins, team status reviews, online dashboards, and the like. This means that when a team member is in trouble, it may take an unacceptably long time to find out. On the other hand, by checking in too frequently with team members, you may be inadvertently telegraphing signs that you don't trust them to follow through on their commitments.

Team leaders who have a habit of micromanagement, however noble their intentions, tend to have difficulty mobilizing, motivating, and energizing their team members. What's more, they tend to create a kind of enforced helplessness among team members. Not only does this require considerably more management time and attention, but it also prolongs the time in which team members can develop self-sufficiency. The frequent outcome is that team members cannot perform to their true capacity, and team leaders become increasingly frustrated at the neediness of their team members.

Some situations certainly call for more check-ins rather than fewer, especially if there's a lot on the line, and you're not completely comfortable that team members are doing what's needed to achieve agreed-upon goals. Other situations will be less clear-cut. For example, a new team member from another country has repeatedly promised some information that would be nice for the team to have right now, but it's probably not essential for their success. You wonder, "Do I check

in yet again, or should I hang back, waiting to see what happens?"
It depends.

Here are some tips to help you decide how best to make sure everyone's on track, on target, on schedule, and on board, without having to become the kind of micromanager that most people prefer to avoid at every opportunity:

- **Gain agreement as to how project updates will be handled.**
  For example, what level of detail is needed, by whom, and by when? Will updates be sent to everyone, to just a few, or will details be posted somewhere instead?
- **Discuss under what conditions are certain delays acceptable, and who needs to be notified when unforeseen circumstances occur.** Will you as team leader always need to be notified ASAP, or are there some cases where a slight delay in notification is acceptable?
- **Create an environment in which it's OK to acknowledge problems or to raise issues that can hamper success.** Make sure team members are not withholding vital project information because they fear retribution or criticism for themselves or their team members, especially in a public forum. If you punish the bearers of bad news, you'll find it harder to get any news at all.
- **Verify that the perceived value of creating project updates justifies the time and effort that may be required to produce and attend to them.** If not, some team members may put dealing with project updates at the bottom of their priority list. For example, do all team members have equal access to the means by which project-related information will be documented and shared? If so, do they actually have the time required to plug in and routinely update their data, and do they have time (or inclination) to read everyone else's?
- **Recognize the importance of an agreed-upon communications plan.** Be sensitive about how team members prefer to provide and receive information. Although some may agree to populate a spreadsheet, for example, they may feel more comfortable offering verbal updates on weekly calls. (See Chapter 5 for more tips on virtual team communications planning.)

- **Be aware of how cultural and language differences may affect the frequency, content, formality, and timeliness of communications.** "A brief status report sent out early in the week" may be interpreted a hundred different ways. Notions of "brief" will certainly vary, as will "early in the week." If specificity and consistency are important, be explicit right up front, and make sure you have a shared understanding.

- **Listen to how people receive your requests for information.** Are their responses increasingly terse? Do they seem pleased or annoyed that you're interested? If the latter, ask yourself why. Better yet, ask them, diplomatically, and encourage them to give you an open honest response. This is best done 1:1 via phone rather than in writing or on a team call.

- **Evaluate whether your requirements for information are realistic or even necessary.** When a team is far flung and you feel a need to keep vigilant watch over their activities, consider the impact your requests for information will have on their morale and workload. Is this request something you have control of, or is this something that the company requires? If the latter, explain the reasoning. People may not like having to get you the needed information any better, but at least they'll appreciate that you took the time to provide a rationale.

- **Be conscious of whether you're assuming the best or the worst about people, and assess how your actions and attitudes reflect your beliefs.** For example, if you believe that most team members are doing all they can to fulfill their commitments, think about what kind of message you're sending by insisting on daily written reports and weekly conference calls. If you know they're drowning in work, acknowledge that fact, express appreciation, and help them to sort priorities if you sense your guidance might be welcome.

- **Be flexible at different intervals.** Some phases of a team, or a given project, require more intense, frequent communications than others. Ask yourself, as well as your team members, at what point it's appropriate to declare that "enough is enough." Question whose purposes are really being served at any given time, and be prepared to modify requests accordingly.

When the pressure is high and all eyes are on your team to pull off some heroic achievements, micromanagement can be a natural tendency. But that management style can be particularly damaging to virtual teams when trust can be so difficult to create and cultivate, and so easy to break. If you can thoughtfully balance your desire to keep tabs on everyone's work with the need for team members to develop the kind of confidence and competence they need to establish self-sufficiency, everyone wins.

8.3 Recognizing and Addressing Signs of Dysfunction to Avoid Irrecoverable Problems Later

If you're part of a virtual team, you learn to develop a sixth sense for knowing when dysfunction has crept in. The signs become clear over time, even though you can't see vital body language or hear side conversations. People start making excuses for missing the weekly conference calls. Or maybe they don't even bother to RSVP. When people do show up, they grunt monosyllabic responses as they pound away on their keyboards. Conversations that take place are often stilted and terse, with little real interaction. When people commit to deliverables, they sometimes renege without warning. When you try to discover what's wrong, members politely refute the notion that anything is awry.

It's tough enough to lead a dysfunctional team when you can see the members and speak openly eye to eye. But when you're leading a virtual team that's become disengaged and dispirited, it takes special skills and approaches to re-engage and motivate those who have drifted away. Here are some practical steps virtual team leaders can take to get team members back on track:

- **State your observations with specific examples and express your concerns.** Start by sending an e-mail with a strongly worded header to implore people to attend the next team meeting. In your header and in the first few lines of your message, state your observations about the team's behavior. (For example, "I have noticed that many people have dropped off our calls. People who attend are not really present. Some of you are reneging on promises, and others are declining to pitch in to help others as you used to. As a result, I feel like the team is falling apart. I am asking that all of you fully participate in our next meeting so that we can explore the real issues and decide what we can do to get ourselves back on track.") Follow up with a phone call to make sure people have read your message and plan to attend.

- **Listen intently.** Once on the call, describe in more detail the kind of behavior you're noticing that's causing the team to deflate. Leave room for silence and reflection. If no one responds after a while, ask for validation. Are they seeing things the same way? Is there something you are not seeing? You may start by acknowledging some of your own issues (e.g., you haven't had enough time to provide thoughtful feedback to everyone lately, or you have a new manager who is pulling you in multiple directions). By sharing your own perspectives, you can encourage others to discuss their own barriers to participation.

- **Create a safe discussion space.** Consider opening up an anonymous conference to precede, augment, or follow up on a realtime phone discussion. Craft some carefully worded questions to elicit honest responses without inhibition. Include a variety of questions such as: on a scale of 1–10, how energized do you feel to be part of this team today? Or: One of the toughest challenges I face in fully participating on this team is ____. Or: One of the greatest rewards of being on this team is ____. Providing an avenue for anonymous written comments may help create a level playing field among different cultural or personality types.

- **Crystallize the underlying problems.** Your team has identified the symptoms that are preventing real collaboration. Now it's time to name the real problems that are causing the behavior. You can try doing this on a team call, especially if your team had demonstrated a high degree of trust for each other, and for you, in the past. You might also try using a web conference tool to enable anonymous input, either synchronously or asynchronously. Alternately, you can ask someone who is not part of your team to interview team members in confidence. After all, if you are perceived to be the problem, members will be reticent to say much on a team call regardless of the vehicle. As a team, set priorities for problem-solving, starting with problems that have the greatest impact on the team's performance.

- **Collaborate on the best solutions.** Once the team has developed a deeper understanding about the dysfunctional behaviors and the underlying problems, it's time to brainstorm solutions. Generate ideas for at least a few quick hits right away, such as arranging a meeting time more convenient to all, or setting up a team portal for easier document sharing. Tackle the tougher issues next, such as reducing the amount of rework required when the organization shifts direction, again, or identifying resources outside the team who can be tapped to help in certain circumstances. Document the related actions and responsible people, and keep the team updated as to your collective progress. Try using a web conference tool to solicit an impressive array of ideas in a short time.

- **Seek commitments to be part of the change.** Elicit from each team member specific actions he will take to lead to more successful collaboration. These commitments should be made to the entire team. For example, a team member who habitually skips the weekly team meeting might make a commitment to be present at least 75% of the time. Or the person who multitasks her way through every call might promise to clear her desk and calendar and be fully present at future meetings. The person who says yes to everything but fails to deliver almost every time might promise to be honest about what he can take on, and will alert everyone ASAP if he was being unrealistic about what he could deliver.

- **Take the team's temperature often.** If a team has been allowed to drift apart for any length of time, it may take many attempts to get them back on course. Find a variety of ways to check in, both as a team and with individuals. Try a simple periodic phone call to each member to check in. Or ask for feedback on a call. You can also try using some sort of online tool that allows all members to provide a few quick responses, the results of which can be later shared with the team. Of course, observing the team's performance against agreed-upon measurements may be the best indicator as to whether the team is collaborating successfully.

- **Reach out to disaffected individuals personally.** Some people will be harder to turn around than others. In some cases, you may have to let go of those whose behaviors are toxic to effective remote collaboration. Take the time and energy to reach out personally with greater frequency for those whose active participation is vital. Make calls, send e-mails, or meet them face to face if possible. Express your concern, lend your support, and provide candid feedback. You want to be honest about your frustration or disappointment, emphasizing the impact on the work of the team. At the same time, you need to praise their knowledge and skills so critical for the team to achieve its goals.

- **Consciously model best practices behavior.** Treat team calls as the most important event on your calendar. Let people know what's expected of them in advance. Come prepared with an agenda and stick to it. Check in to see how people are doing. Project enthusiasm and energy. Applaud team and individual achievements both large and small. Make team meetings engaging and productive. Use technology wisely when it can accelerate results, elicit needed input, or otherwise increase the effectiveness and efficiency of each meeting. Respond promptly to e-mails with insightful information. Admit when you feel you're falling short of expectations, and explain why.

The best way to overcome dysfunctional behavior of remote teams is to nip it in the bud. To do this, you need to develop antennae sensitive enough to alert you to the first sign of trouble. So, if people have stopped exchanging ideas on calls or if team members are ignoring e-mails from others, check in with the team as soon as possible. Be direct about what you're observing, articulate the impact on the team and its shared goals, and declare what you need to have the team do as a result. Diagnosing team problems without benefit of nonverbal behavior is difficult. But applying remedies that can help an ailing remote team get back on track is harder still. Avoid problems altogether by checking in early and often.

8.4 When Your Team Is about to Implode: Watch for Signs, Act Fast

Earlier, we discussed how to develop carefully tuned antennae to sense when things are going off track. In this section, we're looking at a worst-case scenario, where the virtual team leader is blind-sided by a precipitous collapse of a team, where the leader may not have seen the warning signs, or simply did not know how and when to intervene, and with whom.

Many of us who follow the Boston Red Sox may prefer to forget the Great Collapse of 2011, when they went into an unstoppable tailspin in the final month of the season, erasing what many thought to be an unconquerable lead, losing a playoff berth in the final inning of the final game of regular season play.

Although it's true that baseball is only a game and that the 2011 Red Sox were just an overpaid underperforming group of players, I wanted to salvage something positive about their shocking demise. Specifically, I wanted to get a better grip on how and why a talented skilled group of players can suddenly stumble into oblivion. (Notice, please, that I did not refer to the 2011 Red Sox as a team. They were a collection of individuals who each seemed to speak a different language, play by his own set of rules, and work toward his own goals. There was no apparent chemistry, cohesion, or collaboration that are the hallmarks of truly great teams.)

All of us can learn from the following checklist of contributing factors. For virtual team leaders, the underlying reasons for underperformance may be much harder to root out, and the interventions more tricky to apply. The key is to act as soon as the first red flag goes up, rather than hoping the problems will just go away on their own.

- **Creating a team culture.** When new players join the team, they need help becoming immersed in the "local culture." For example, each team has its own principles and norms about socializing, practicing, public behavior, surfacing issues, and resolving conflicts. Team leaders need to make sure that new players become assimilated as quickly as possible into the prevailing team culture, which might mean assigning a "buddy" or two to shepherd them along in the early days. "Ultimately, you don't need a team that wants to go out to dinner together, but you need to have a team that wants to protect each other on the field and be fiercely loyal to each other," said now-former Sox manager Terry Francona. "That's what ultimately is really important."

- **Seeking superficial harmony instead of facing conflict head-on.** Some managers try to bolster spirits in the clubhouse by giving players positive strokes. Confronting underlying issues directly and encouraging players to speak plainly about their own performance as well as the team's allows players to hash out their differences and shift destructive behaviors. Avoiding conflict and tip-toeing around tough issues may feel like the "safe" thing to do, but it's actually one of the surest ways to accelerate a team's demise once it's started going in that direction.

- **People are not operating from their real strengths.** Like a sports team where some excel at defense or offense or speed versus strength, each team member brings certain gifts, experience, talents, and expertise. Add to that, some players devote more time to honing their skills than others. Leaders need to provide an environment where people can move out of their comfort zones, stretch themselves, and excel in brand new areas, rather than performing merely competently by executing the same level of performance that they have done for years. Even if you have a team member who is capable of greatness, if that person does not exploit her potential, her contributions will be no greater than, and perhaps even less than, some of your inexperienced staffers.

- **Avoiding accountability.** When team members aren't accountable for their own actions, they hurt their own performance. But when they duck responsibility for calling out other team members for their behavior, the performance of the whole team suffers. For example, if a right fielder knocks over the center fielder when both try to catch a fly ball, they both quickly need to agree on a ground rule to prevent catastrophe the next time. Similarly, if you have a team norm that says people who don't do prework must get caught up on their own time, everyone needs to call out the transgression when someone interrupts to ask a question that was covered in the prework. A close-knit team knows how to hold each other accountable without rancor.

- **Egos getting in the way.** We may cheer when our team outbids a competitor for a sought-after superstar. But when the celebration's over, we know that stardom often brings with it big egos, the kind that can divide a team and breed resentment. When a team has multiple stars, whether it's the deified athlete, a social marketing whiz kid, or the fair-haired child of the CEO, big egos can lead to problems that can escalate quickly. Regardless of the quality and depth of stardom a team may possess, leaders need to be scrupulous about requiring that everyone be treated fairly, playing by the same rules, and working toward the same goals, regardless of stature or salary. Otherwise, power struggles can quickly deplete energy and derail a team.

- **You've got the wrong players.** Half the team might have exceptional skills in certain areas, whereas other talents are notably lacking. Some players may have been inherited, some were invited to join based on past successes, and others may have been picked up because no one else claimed them. In addition to the skills each brings to bear, consider the extent to which some people like to collaborate and others want only to pursue individual goals. If you can't find a way to coax and cultivate your team's talent to make the best use of what they have, and to make up for what they don't, you'll have to decide who stays and who goes. The longer you put off the decision, the harder it will be to pull together as a team.

- **Magical thinking.** We all fall prey to irrational optimism, especially when we have run out of ideas about how to turn things around. ("If only we can get through this series, we're sure to go the rest of the way"; or, "We can overcome our project delays if we all work hard and do our best.") Really? You think so? This kind of unrealistic hopefulness, although it might act as a salve at the moment, can actually disable us from taking any real action to get the team back on course. It's true that hope can act as a powerful motivator. But hope without an accompanying plan for change can keep a team mired in the muck, unless of course you happen to have the benefit of a magic wand.

- **Negative thinking.** It's one thing to be aware of your limitations and have a plan to work around them, but it's another to become hobbled by self-doubt. Anyone who watched the Sox play the last games of the season saw a group of men who went through the motions, spirits lagging, while the opposing teams exploited their weaknesses, which had far less to do with physical problems than collective despair. Remind team members of the brilliance they're capable of and explore ways they can regain their lost luster. Well-placed honest feedback can do wonders to turn a team around, but constant haranguing will lead to a pervasive negativity that can't easily be overcome.

- **It's just no fun anymore.** When people dread going to work, they simply won't be operating at peak performance. No one expected the 2004 Red Sox to win their first world championship in 86 years. Throughout the playoffs, they stayed loose, joking and laughing both on and off the field. As outfielder Johnny Damon said, "We are just a bunch of idiots, having a great time." It's no coincidence that they became the only team in baseball history to come from three games behind to win a seven-game playoff series. When people have fun and like their work, unhampered by pressure, they aspire to perform their best. Find ways to make work fun, especially when the pressure is on.

Even the mightiest team can fall apart without apparent warning. Leaders need to be vigilant about looking for those small telltale cracks before they become irrevocable fissures that can tear a team apart. As (now former) Red Sox general manager Theo Epstein said after his team's epic collapse:

> When you go through what we just went through, you can't look past anything. You have to take a hard look at every aspect of the organization, one's self included, and ask, "Is this exactly the way we want it to be? If everything is going right … if we're exactly who we want to be, is this element of the organization functioning the way we want it to?" If the answer's no, then we have to go out and fix it. And that's going to be a very difficult, very painful, painstaking, thorough process. But the bottom line is we failed.

For virtual leaders the signs of disaffection and dysfunction can be harder to detect. But at the first sign of discord, whether it's a sarcastic tone, an angry word, or a passive-aggressive response, it's time to act quickly and decisively. Worst case: you may be overreacting to a momentary lapse. Best case: you may have helped the team avoid an irretrievable meltdown with just a few thoughtful interventions.

8.5 How to Disengage Your Virtual Team in 10 Easy Steps

One major challenge that comes up in just about all of my virtual leadership coaching sessions is how to keep virtual team members engaged, enthusiastic, motivated, and energized. Rather than coming up with a list of practical tips to keep virtual team members engaged, I decided to create this list of sure-fire steps virtual leaders can take to *disengage* members. If you spot any of your behaviors here, think about what you can do differently to flip it around and keep team members engaged.

1. **Allocate tasks that encourage independence.** The less dependent people are on others, the more likely they'll be to get their work done on time. Structure assignments so people can complete their work on their own. You know how some of these people just love to chat! Don't give them any more reasons than you have to. Once they get on the horn, there's no telling how much time they might fritter away.

2. **Keep goals fuzzy for greater flexibility.** You never want to be too explicit about team goals, in case you need to change them in a hurry. Better to give out a slew of tasks and deadlines conveying the appropriate sense of urgency, keeping people so focused on their deliverables that they won't have time to figure out how their contributions fit together toward achieving group goals (if in fact, by some miracle, they do). After all, it's much easier to check off items on a task list than stepping back to see how actions support overall goals.

3. **Don't bother with team norms.** People will just ignore them anyway. And you'll have the unenviable job of insisting that everyone live by them. There's no way you can stop John from multitasking on team calls, or Mary from sending hourly e-mails to everyone on the team, or Max from criticizing everything people have to say, so why even bother? Better to ignore dysfunctional behavior and hope it goes away of its

own accord, or better yet, hope that someone else on the team takes care of it for you.

4. **Check in with team members early and often.** Don't waste time asking about how people are doing or what you can help with. Cut to the chase and ask them when they'll be done with their latest assignment. After all, you can't see how they're spending their days, and you want to make sure they're focused in all the right places. Use IM, e-mail, phone, text, and anything else you can think of to make sure they know you're concerned about them (and the state of their projects). Some may call this micromanaging, but you know you're just keeping tabs.

5. **Dole out important information to certain people first.** Start spreading the news with people to whom you're closest. Grab a cup of coffee so you can give them the low-down and hear what they have to say. Catch the others up when you have time, maybe at next week's team meeting, or in your Monday morning team e-mail. They'll probably find out the news from other people first, anyway. They can't really expect you to take the time to call each one, can they?

6. **Emphasize efficiency and brevity on team calls.** Keep meetings super-quick by discouraging questions and dismissing divergent perspectives. Who's got time for a real conversation? Send out overly crammed agendas in advance, so no one will be tempted to bring up any out-of-bound topics. If people insist on bringing up issues or questions that you don't want to cover, encourage them to send an e-mail instead. (Of course, you'll never have time to respond, but they'll feel better getting it down in writing.)

7. **Economize your communications.** Your days are busy enough getting your own work done and catching up with people who work nearest you. How on earth can your remote team members expect you to be spending so much time with them? Instead of getting caught up in time-consuming calls, send e-mails or IMs. In extreme cases, especially where performance is lagging, set up a 1:1 meeting to provide needed coaching, but make sure they know this will be the exception and not the rule.

8. **Cancel unnecessary meetings.** Most of us need more meetings like a hole in the head, right? If you have team meetings

or 1:1s scheduled, try using e-mail or IMs instead. As a courtesy, cancel meetings at least an hour in advance, and let participants know you expect them to use this gift of time productively. Invite them to call you if they have something urgent to discuss. Use caller ID to ward off questions that are likely to be a big time sink for you.

9. **Let the big talkers take over**. Think of all the work you can get done while someone else takes over your team meeting. If some people keep quiet, they either just don't have anything to say, or maybe they're multitasking, too. You don't want to pull teeth to force people to speak. It takes way too much energy. Plus, insisting on hearing from everyone will soak up too much time. Consider it a blessing that only a couple of people ever say anything.

10. **Take advantage of always-on technology.** Every team member has a smartphone, courtesy of the company, so it's only fair you insist that they use it to get work done. Any time. From anywhere. With a smartphone, it's easier than ever for people to slip away from a family dinner, soccer game, or even a vacation, to access an important document, join a team call, or answer an urgent question. Insist that people be accessible via IM or text around the clock, just in case. After all, working flexible hours goes both ways.

There are hundreds of ways, large and small, in which virtual team leaders can alienate their employees without trying very hard. If you want to engage your team members, take a few of these tips, and try the converse.

8.6 Summary

Even the strongest teams lose their way once in a while. So do the most talented team leaders. The successful virtual leader must be vigilant about detecting signs of trouble early (and sometimes often), paying attention to even the most seemingly innocuous telltale signs. At the first hint, act quickly and decisively to validate issues, discover root causes, and then collaborate with your team members to create solutions that help get people back on track.

9

SPECIAL CHALLENGES
OF FACILITATING
VIRTUAL MEETINGS

First, let's define what we mean by virtual meeting. To start off, just because some or all participants are remote, a virtual meeting is no less real than a meeting where everyone participates face to face. So virtual meeting = real meeting. As such, it deserves every bit of thoughtful planning and masterful facilitation as a face-to-face (FTF) meeting. Even if only one person attends remotely, and all others participate via speakerphone from a conference room, we consider this to be a kind of virtual meeting that must be designed and facilitated differently from one where all participants are located together.

"Hybrid" meetings are those where some participants attend from one location and others join remotely. These can be the most challenging kind of meetings to get right, as you have to work hard to create a level playing field when some participants are perceived to have potential advantages associated with being physically together.

Virtual meetings can be same time (synchronous) or any time (asynchronous), or a combination of the two. For example, participants might cover the first agenda item by typing into an asynchronous online conference area, and then convene via phone and web meeting sometime later, to cover the rest of the agenda. For this book, "virtual" and "remote" meetings mean the same thing.

A final definition: a meeting is a formal structured event that typically has a defined purpose, intended outcomes, agenda, particular participants, and a meeting leader. This is in contrast to chat forums, social media hangouts, wikis, microblogging feeds, e-mails, IMs, and the dozens of other ways virtual team members can communicate outside of regularly scheduled meetings.

9.1 What Makes Virtual Meetings So Challenging?

In virtual meetings, when some or all people can't see each other, the meeting leader faces a host of challenges when trying to keep conversations focused and people engaged. For starters, people can tune out invisibly and without notice, leaving the meeting leader to wonder, "Are people listening? Are they interested in the topic? Do they agree or oppose? Are they upset? Bored? Are they all doing e-mail? Is this meeting worth having right now?"

Next, in the absence of nonverbal cues, we have a hard time getting an accurate read on the meaning of one's tone, cadence, silence, snickering, throat-clearing, or other audible cues. If we see people folding their arms and rolling their eyes, we have a pretty good idea how they're feeling. When we go only by our ability to listen, we have to work much harder to decipher cues, which often lend themselves to misinterpretation.

Because virtual meetings must be kept brief to accommodate ever-shortening attention spans, meeting leaders are pressured to achieve objectives in a relatively short time. (And if you think that a half-day FTF meeting can be replaced by a three-hour virtual meeting, think again!) This means that everyone must come prepared to jump into a productive conversation from the first moment, which almost always requires that some kind of prework be done by all.

Cultural differences can be tough enough to navigate through, even when people do have nonverbal cues to go by. In virtual meetings, when our powers of observation are limited to our ears alone, the opportunities for misunderstandings and frustration due to cultural differences are exponentially greater. When participants speak different native languages, meeting leaders need to allocate more time for conversations. They also need to consider whether they'll need to modify their facilitation techniques for a multicultural, multilingual audience. When global teams span several time zones, finding a meeting time that's convenient for all can be tricky.

Given that some people regard virtual meetings as "not quite real," some teams tend to be more casual about follow-up, including taking and sending meeting notes, following through on actions, and making good on commitments. The frequent result is that much time is wasted in the subsequent meeting to rehash decisions, actions, and next steps, especially if more than a week or two has elapsed since the prior meeting.

9.2 Six Critical Factors for Running Productive Virtual Meetings

My colleague, Julia Young at Facilitate.com, says it best, "The goal of any successful virtual meeting leader is to facilitate an engaging conversation around a focused agenda with only the necessary people who are prepared to accomplish a clear set of outcomes."

Sounds pretty straightforward, but it can be very tough to do without a great deal of practice. Julia's "Six Critical Success Factors" model (shown in Figure 9.1 and detailed in Table 9.1) provides a logical, practical framework for thinking about how team leaders can achieve this goal through a virtual meeting (or series of meetings). The centerpiece of this model, keeping participants engaged, is the single most challenging aspect of planning and running a productive virtual meeting, and it's the element that all the other success factors must support.

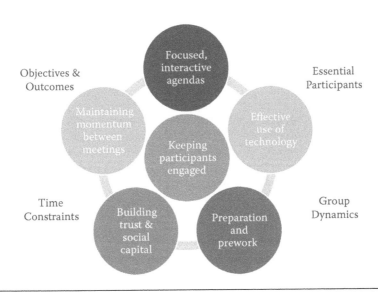

Figure 9.1   Six critical success factors for virtual meetings. (© Facilitate.com. Reprinted with permission. All rights reserved.)

Table 9.1   Six Critical Success Factors for Virtual Meetings

- **Designing a series of focused agendas.** Effective virtual collaboration tends to be done through a series of short virtual meetings, up to 90 minutes at a time. Prework (i.e., work done by participants prior to the meeting) is often an essential tool to limit actual meeting time and ensure that the precious "realtime" spent together is focused and productive. As we plan our virtual meetings, it is important to determine the level of interaction needed to meet our objectives, whether they be information sharing, data gathering, idea generation, problem solving, or decision making. The type and level of interaction needed for each part of our meeting will, in turn, determine the tools and technology we need to use. Organizing a virtual team's work into a series of short meetings enables us to involve only the people necessary at each stage. Large virtual meetings can be unwieldy and tend toward one-way communication rather than meaningful interaction and collaboration. A series of short focused meetings with good communication in between allows us to involve the right mix of people at the right time.

- **Effective use of technology.** There are many different kinds of technology available today for augmenting teleconferences with tools to increase participant engagement and interaction. As virtual meeting leaders, we need to be aware of the full range of technology tools available to us and become practiced and proficient in their use. There is no shortcut here; we need to do the research and hone our skills. We also need to make sure all participants are equally confident and comfortable in the use of virtual meeting tools. Low-tech options can work equally well in a virtual meeting as in a FTF meeting. Use a simple handout if that is all that is needed. First and foremost, we need to create the right environment for a productive conversation.

- **Preparation and prework.** There are two important reasons to design prework into our virtual meetings. The first is to prepare participants to take full advantage of the session by thinking ahead about the content, formulating ideas, or getting to know others in the group. Participants who have completed well-thought-out prework assignments are "primed" for active and open participation in the realtime event. The second is to get ourselves ready to run the session effectively. By knowing more about our participants and their interests, we are in a position to develop focused questions that will stimulate ideas and keep participants engaged.

- **Building trust and social capital.** Trust and a comfortable level of social or personal interaction are key ingredients of effective virtual meetings. As meeting leaders, we lose some of our influence in managing the team dynamics in the absence of physical presence, and need to find new ways to create a healthy dynamic within the group. We also need to help create a level playing field among all meeting participants, and need to find ways to cultivate trust for ourselves and across the team.

- **Maintaining momentum between meetings.** Transitioning from a FTF event to a series of short virtual meetings means that we have to work harder to maintain momentum and motivation between meetings. Meeting leaders must wear many hats at once: project manager, logistics coordinator, and the social glue that holds the group together. Although meeting leaders need not take on the full responsibility for maintaining momentum, we do need to recognize the importance of continued communication and engagement and devise ways to keep in touch with participants and keep them in touch with each other.

(Continued)

Table 9.1 Six Critical Success Factors for Virtual Meetings

• **Keeping participants engaged during the virtual meeting.** We all know how difficult it is to stop people from multitasking in a virtual meeting. (Who among us is not guilty of this ourselves?) As meeting leaders, our ability to keep participants engaged starts well before everyone is on the call. Each of our critical success factors contributes to maintaining participant engagement: if we design a short focused meeting, select the right technology, invite the right people, prepare participants well, attend to the group dynamics by building trust and social capital, and create momentum with good communications, we're far more likely to keep people engaged.

Source: © Facilitate.com. Reprinted with permission. All rights reserved.

9.3 Summary

In the next few chapters, we explore these success factors through the series of tips that cover planning, design, and facilitation of successful virtual meetings in much more depth. Designing and planning a great virtual meeting may not be rocket science, but it takes a surprising amount of thought, trial and error, and plenty of practice to get it right.

10

ABCs of Designing
Great Virtual Meetings

Bringing people together for face-to-face (FTF) meetings can be costly, in terms of time, money, and missed opportunities when people have to put their "real work" on hold for extended periods of time. Some people insist that there's no acceptable substitute for eye-to-eye contact with people in the same room, especially if a key goal is to build new relationships or mend broken fences. And although that may be true in some cases, virtual meetings have become the norm, especially for global organizations.

That's why today's virtual leaders need to "crack the code" for designing, planning, and leading virtual meetings that keep people energized, actively participating, and focused on the conversation at hand.

In this chapter, we focus on some of the most critical variables to consider when designing and planning a great virtual meeting. Make no mistake: planning and leading virtual meetings that are consistently productive and engaging takes creativity, thoughtful planning, and a lot of practice. We include tips to design "hybrid" meetings, where some people are physically together and others are remote, which are especially tough to do well.

First, a word about knowing when to push for FTF meetings over virtual conversations. Certainly, there are situations when meeting face to face will be crucial to achieving goals, especially when a high degree of trust is required among certain members. Consider which relationships are most important to the team's overall success, as well as the quality of existing relationships. In cases where people are heavily dependent on each other for their success and where no trusting relationships have been created, meeting face to face may be essential. Or when a subset of the team needs to collaborate intensively to accomplish a great deal of work in a short time, investing in a FTF meeting can have a huge payoff.

If you are wondering whether you can have a successful meeting by meeting remotely versus speaking eye to eye, Table 10.1 provides a checklist to help you assess the likelihood of success. If you agree with most of the statements in the first part of the table, chances are a FTF meeting will help you achieve your objectives better than a remote meeting. If, however, your responses suggest that a remote meeting may be just as effective, consider the questions in the second part of the table as you make your plans. If you answer "yes" to most questions in the second part, then with good planning and clear communications, chances are you will have a successful remote meeting.

When requesting a FTF meeting, be sure to make a persuasive business case, quantifying expected results both with and without the meeting. Also consider how videoconferencing can help achieve some of these objectives, especially if you have access to high-quality "telepresence" videoconferencing solutions where participants feel as though they are all physically present. Not all videoconferencing capabilities are created equal, and some systems may actually interfere with collaboration, rather than enabling it.

10.1 Creating a Realistic Agenda for a Productive Virtual Meeting

When planning a virtual meeting agenda, a few rules almost always apply:

- **Aim to go no more than 90 minutes**, even when you're convinced that your meeting will be completely captivating for all participants. Two hours is the maximum recommended virtual meeting time, and only if you absolutely must.
- **Because most people are hard-wired to multitask these days (even meeting leaders!), think about how you can design a meeting where "multitasking on task" plays a central role.** That is, capitalize on participants' predilection for juggling multiple tasks by embedding various kinds of multitasking into the conversation. For example, I like to set up online flipcharts for brainstorming ideas or problem-solving solutions. Participants can type in responses within a few minutes, and then we can discuss ideas as a group. Thus, I am giving participants something useful and topical to do with their itchy fingers, which might otherwise be typing up e-mails or surfing the web.

Table 10.1 Meeting Face to Face or Remotely: Evaluating the Options

	STRONGLY AGREE	AGREE	NEUTRAL	DISAGREE	STRONGLY DISAGREE
1. If we achieve our intended outcomes, we have a lot to gain. If we don't, we have a lot to lose. It's critical that we're successful.					
2. We have a great sense of urgency to achieve our goals. Time is of the essence. Delays are unacceptable.					
3. A high degree of trust among team members is critical if we are to meet our objectives.					
4. In-depth conversations are necessary for us to make well-informed decisions and reach agreement.					
5. Topics of discussion are likely to be contentious or may cause conflict or evoke emotion.					
6. Tapping the enthusiasm and energy of all participants will be important for us to achieve our goals.					
7. It is unlikely we can achieve our objectives over a series of several brief meetings.					
8. Creative brainstorming and problem-solving will occupy much of our meeting time.					
9. It's critical that we test understanding, validate assumptions, and clarify expectations.					
10. Key participants represent a variety of cultures and time zones.					
11. Not all participants have equal access or comfort with technology used for meeting remotely.					
12. The cost for failing to achieve our objectives is likely to exceed the costs for assembling participants face to face.					

(Continued)

Table 10.1 (*Continued*)   Meeting Face to Face or Remotely: Evaluating the Options

PLANNING FOR A SUCCESSFUL REMOTE MEETING

1. Are participants likely to stay focused on the work if we meet remotely?	Yes	No
2. Do we have access to facilitators who are skilled in planning and running remote meetings?	Yes	No
3. Is it possible to carve up the agenda into smaller "chunks" and still achieve the desired results?	Yes	No
4. Can some of the work we would ordinarily do face to face be accomplished either before or after the meeting instead?	Yes	No
5. Are participants likely to pay attention to prework and advance reading that may be necessary if we run the meeting remotely?	Yes	No
6. Do most participants know each other and work well together?	Yes	No
7. Are most participants accustomed to working in a distributed fashion already?	Yes	No
8. Have we considered ways to follow up, maintain momentum, and track progress remotely?	Yes	No

Source: Created by Nancy Settle-Murphy of Guided Insights, and Penny Pullan of Making Projects Work Ltd. Reprinted with permission.

- **Plan for 80% active participation and no more than 20% passive participation.** That is, design your virtual meeting so that people are actively engaged in conversation about 4/5 of the time, and in listening mode only 1/5 of the time. (Yes, that means posting those lengthy presentations or documents for review in advance of your meeting, leaving precious meeting time for active conversations vs. one-way monologues.)
- **Assume some kind of prework or prep will be needed by all participants.** Virtual meetings have to be kept brief to ward away the tendency to multitask; therefore, we need to make sure that all participants come to the virtual table ready, willing, and able to contribute from the first minute. This means that all participants need more or less the same level of knowledge or background information about the topic at hand. This almost always means that some type of prereading or prework is required.

Here are more tips to keep in mind when creating your virtual meeting agenda:

- **Break down your objectives into manageable bites.** Most meetings have multiple objectives. Say, for example, you had planned a three-day project team kickoff meeting intended to build new relationships among participants, reach agreement on the project scope, clarify roles and responsibilities, create a shared team communications plan, brainstorm the top issues facing your division, agree on priorities, and map out an action plan. Each of these objectives can probably be met in ways that do not require any FTF meetings by all (or most) team members. Create a three-column table with objectives down the left side, one per row. Next to each objective, list participants who need to be involved in the middle column, signifying the nature of their involvement (e.g., provider of input, decision maker, idea generator, approver, etc.). Finally, in the right column, list possible ways to achieve each objective. Compartmentalization of meeting objectives is the first step to designing workable virtual options.
- **Be realistic about what conversations need to take place, and by whom.** You can't expect to cram an agenda meant for a three-day FTF meeting into a two-hour virtual meeting.

Consider which objectives require real-time conversations and which can be achieved in other ways, such as via e-mail or chat forums. Also think through the type of conversations that are needed. For example, are healthy debates required to reach a decision? Do people need to spend time brainstorming a solution? Some conversations take longer than others. Make sure you aren't trying to accomplish too much.

- **Work with the time you really have.** Consider time across several dimensions. First, how soon does the team need to deliver results? Is two weeks from now too late? Will partial results within a week be acceptable? Second, how much time are participants willing to give? Will they allocate time for preparation and participation in three 90-minute working sessions over the next 10 days, or are they more likely to set aside time over the course of a single day? As you divide objectives into manageable chunks, consider how much total time you're asking of each participant, and set expectations right up front.

- **Create and communicate clear objectives, both in advance of the meeting and at the start.** Test for shared understanding to make sure you're all in sync. Allocate sufficient time to do this, especially if the topics are likely to be complex, contentious, or sensitive. Although you may accept modifications, resist attempts to overhaul or add to objectives. After all, you've created an agenda and invited exactly the right people for the original objectives. "Park" any other objectives for the end of the meeting, when you can agree how best to handle them. Be prepared to refer to objectives frequently as a way to keep conversations focused.

See Table 10.2 for a template of prework communications.

10.2 Selecting the Right Participants

One reason so many virtual meeting participants tend to multitask is that they really don't need to be there. (And who can be faulted for multitasking when most of the meeting is irrelevant and uninteresting to them?)

Table 10.2 Virtual Meeting Prework Communications Plan Template

MEETING:

DATE/TIME:

TELECONFERENCE # AND PASSCODE:

ONLINE ACCESS: HTTP://___

TIMELINE		COMMUNICATION	FORMAT (E-MAIL, PHONE, IN PERSON)
	A	Announcement/invitation to attend Purpose and outcomes Date and time Who should attend/why you should attend Expectations for participation including prework Requests for response Contact information for questions	
	B	Prework instructions Reading attachments Online prework steps and log-on information Connections with other participants Deadlines Requests for confirmation	
	C	Calendar appointment Date and time Reminder alert for meeting and prework Teleconference access information Online meeting tools log-on information	
	D	Meeting reminder Purpose and outcomes Date and time Agenda Ground rules Request for confirmation/cancellation Contact information for questions	

PARTICIPANT	A	B	C	D	PARTICIPANT	A	B	C	D
	◆	◆	◆	◆		◆	◆	◆	◆
	◆	◆	◆	◆		◆	◆	◆	◆
	◆	◆	◆	◆		◆	◆	◆	◆
	◆	◆	◆	◆		◆	◆	◆	◆
	◆	◆	◆	◆		◆	◆	◆	◆
	◆	◆	◆	◆		◆	◆	◆	◆
	◆	◆	◆	◆		◆	◆	◆	◆

Source: © Facilitate.com. Reprinted with permission. All rights reserved.

Another reason to think carefully about who needs to participate, and when, is that it's nearly impossible to have an interactive conversation with more than a few people at a time. For many team leaders, it's easier just to invite everyone than it is to think through which people really need to be there. In addition, team leaders are reluctant to convey a message to some team members that their participation is not wanted or needed. So rather than providing a rationale for selecting some participants and not others, many team leaders take the easy way out and open all meetings to everyone, which can have many unintended consequences, and few of them positive.

Here are some tips for selecting the right participants for the kind of conversation your team needs to have:

- **Consider who really needs to be involved and in what way.** Does the entire team really need to be involved in every single activity or conversation? Or is it possible for a subset of people to tackle some of these objectives, such as brainstorming solutions to top issues or mapping out a section of the overall action plan? Ask yourself who needs to provide input or feedback before or after a meeting, and who really needs to be involved in the actual conversation at the same time. For example, people can post ideas in a virtual conference room, on a team blog, or even via e-mail in advance of, or after, a real-time meeting. Interviews can also help unearth valuable perspectives in advance or afterwards.

- **Try having some people join for just the portion of the conversation that most directly affects them**, as long as partial participation does not disrupt the flow of the entire meeting. In any event, let everyone know how much you appreciate how valuable their time is, so some don't feel put off when they are not invited.

- **Consider time zones, native language, and relationship to other team members** when determining who should be involved in which conversation. In many cases, conversations among a subgroup can be an effective way to make rapid progress.

- **Make sure the meeting is relevant for everyone.** Maybe some people can join only at critical junctures, freeing them to focus on other pressing work. Perhaps not everybody needs to participate, or can catch up on decisions via a postmeeting summary. The smaller the group, the more that people will feel their voices are really heard, giving them a greater incentive to pay attention. As the number of people who participate on the periphery goes up, so does the likelihood of their "wandering off" to tend to other things.

10.3 "Rightsizing" Your Virtual Meeting Depending on Objectives and Group Size

The design of your virtual meeting will rely heavily on a number of variables. A key factor is the number of participants. For example, you can't expect to have a truly interactive conversation with 150 people, or even 20 people, especially if you are relying solely on an audioconference to connect those participants in real time. You can, however, expect to have rich conversations if you keep the number of active participants to 10 or fewer. See Table 10.3.

On the flip side, if your virtual meeting needs to involve dozens or hundreds of participants, such as when a major new organizational change will be presented, then your virtual meeting needs to somehow accommodate Q&As without disrupting the flow of the real-time presentation.

10.4 Selecting the Right Combination of Technology Tools for the Highest Level of Engagement

If your organization is like most others, you have a host of virtual collaboration tools at your disposal. For example, you probably have at least one web meeting tool that most people tend to use. (Or you may have one "sanctioned" web meeting tool, and others that people might bring in from time to time, with or without IT approval.) Some of your virtual collaboration tools may allow for asynchronous participation, such as chat forums or web portals, whereas others might be better suited to real-time communications.

How best to combine technology tools for the most engaging virtual meeting experience? That depends on a host of factors, starting with your objectives. To what degree is real-time interaction important? How many people do we need to hear from, and at what point? How can we use asynchronous participation for those in different

Table 10.3   Tailoring the Interactive Experience to the Size of Your Group

A virtual meeting gives you the opportunity to increase the number of people who can participate. Your meeting objectives will determine the optimal group size. In turn, the group size will affect your meeting design. In general, the smaller the group is, the greater the opportunity for personal interaction and sharing among the participants, and the larger the group, the less personal the experience.

Creative use of technology and facilitation techniques, however, can maximize the opportunities for interactivity even in larger groups. An ongoing series of project meetings or team conference calls offer more opportunities for connection and depth than a one-time event.

- **Mini-meetings/webinars: 5–10 people.** These are characterized by a conversational tone and the feeling of sitting around a table with everyone having airtime. There is the opportunity to get to know each other and build social capital that can lead to sharing of personal stories and experiences in a trustworthy environment. Many FTF activities are adaptable to this size group.
- **Small meetings/webinars: 10–25 people.** These are characterized by limited airtime for all participants. Web collaboration tools allow everyone to get ideas down quickly on a shared online flipchart to stimulate and focus the discussion. Use voting tools to facilitate the quick collection of priorities, help groups move toward consensus, and confirm agreement with decisions. Share materials ahead of time for more interactive discussions.
- **Medium-sized meetings/webinars: 25–50 people.** Here, the connection with and between participants is more distant and less personal. Web collaboration tools are critical for a high level of interaction and to keep people engaged. Guest speakers help focus discussion and create interesting debriefings on group input. Meetings and webinars of this size require tightly facilitated Q&A; online chat or brainstorming tools can capture questions for a moderator to paraphrase.
- **Large meetings/webinars: 50–150 people.** Panel discussions keep audio conversation lively while collecting comments back and forth between participants on a shared online flipchart. Deliberative polling (pre- and post-) focuses participant attention on the key issues and illustrates changes in ideas over the course of the webinar. Capture group comments online for a documented takeaway. Use teleconference services for breakout group discussions.

Source: © Facilitate.com. Reprinted with permission. All rights reserved.

time zones? How can we set up prework in such a way that all can respond to and view questions in advance, before the actual meeting, to make for a more productive real-time meeting?

Here are some guidelines to help you narrow down your choices. Regardless of the specific tools you use, consider only those tools that are accessible to all participants, and comfortable for them to use. Depending on your goals and budget, you may also consider buying or licensing a tool for a particular meeting.

- **Make it easy for people to participate at a time and place convenient for them.** One of the great benefits of meeting virtually is that more people can participate with relatively little effort and at practically no cost. Consider how you can use asynchronous means, such as setting up a virtual conference room with electronic flipcharts, to hear input from those who won't be participating via conference call, given time zone differences, language barriers, or role. You'll save precious phone time by coming to the virtual table with top issues or proposed solutions in hand for discussion. You can use the same virtual conference room later on by inviting feedback, posting minutes or other meeting documents, or asking people to build on ideas generated during the call. Use the same web meeting tool for the actual meeting as you do for any asynchronous work, so people have had a chance to try it on their own before the call.

- **Provide multiple communication paths to generate more ideas from more people.** Using the telephone alone places unnecessary constraints on a group's ability to generate ideas or solve problems, especially given how brief a virtual meeting must be to keep people actively engaged. Try using web meeting or conferencing tools that enable people to enter ideas for all to see and build on during verbal discussions. Tools that allow participants to prioritize and vote are especially helpful when time is of the essence.

- **Enabling anonymous contributions can help reduce barriers to participation**, especially if topics are contentious or participants are reticent to speak openly due to fear, shyness, or language differences. Find out what meeting tools your organization already has in place, and determine whether they have the capabilities you need. If not, such tools can be licensed, purchased, or rented per event. Some require downloads and others simply require Internet access. When evaluating the cost, consider how much time you may be able to save, given the hourly cost of participants' time, as well the value that can be gained by accelerating results.

- **Use meeting technology only when it can accelerate and improve results**. In some cases, an audioconference alone might be sufficient, especially where the key objective involves relaying information to others. However, if you must solicit multiple opinions, generate new ideas or make complex decisions, using a web collaboration technology in conjunction with simultaneous audio can produce dramatic results. In general, the more you can design your session to take advantage of the right technology, the less time you'll need. Know when the use of certain kinds of tech tools are overkill and may actually inhibit the kind of free-flowing conversation your team may need.

- **Create a level technology playing field**. If you're using any tech tools in addition to the phone, make sure that everyone is proficient in the use of given functions. Getting everyone up to speed at the start of a call can eat up several precious minutes very quickly. Have people practice the tool first in a low-risk way. For example, set up an online conference in advance of the meeting, where people have a chance to experiment on their own time. Invite people to call in 10 minutes early if they want some assisted practice before the meeting begins. Above all, refrain from using all possible bells and whistles, unless there's good reason, because it can be confusing for others and tough for you to manage, unless you're lucky enough to have a production assistant.

- **Be sure to have a back-up plan in the event of glitches**. For example, you may want to push slides and documents out in advance of the meeting in case people are unable to access and download during the meeting. (This can happen for just one or two people, or for everyone.) If bandwidth differences enable some people to fully contribute and others to be delayed or shut out, consider what workarounds are viable. In an extreme case, you may need to reschedule the meeting if the technical problem appears to be temporary. Or you might try reading some content out loud for those who can't easily access the real-time content as it's created, or ask them to voice their opinions, and offer to type them into the virtual meeting area. (See Chapter 12 for virtual meeting troubleshooting tips.)

Technology can be a virtual meeting planner's best friend and, at times, our worst nightmare. Before you choose your set of collaboration and meeting tools, be clear as to how each tool can enable and support participants in achieving meeting goals. Design the meeting according to your key objectives, and then select the combination of technologies that can best support your goals. Don't let the capabilities and functions of your web meeting tools constrain your ability to achieve your meeting objectives. Instead, start with your objectives and figure out how and when to use certain tools accordingly.

10.5 Making Prework Work

Virtual meetings, whether via videoconference, conference call, web meeting, or some combination, must remain focused and relatively brief if you hope to keep everyone from diverting energy and attention to e-mail or other fire drills that compete for attention. To ensure a sharply focused meeting, people must come fully prepared to contribute from the first minute. That means sharing and reviewing documents, including slide sets, in advance. (After all, nothing shouts "time to read e-mail" like a long-winded presentation!)

Here are some tips for helping ensure that everyone comes ready to fully contribute at your next virtual meeting:

- **Give people a reasonable amount of time to prepare**. Establish conventions and ground rules to support thorough preparation by all. Examples include: preparation work must be able to be completed in no more than 90 minutes. Documents must be available for review at least three business days prior to the call. All documents will be available for viewing on the team's SharePoint directory in a place clearly labeled with the meeting date. Participants who enter the meeting unprepared will have to catch up on their own time. Make sure to enforce ground rules equally for all, regardless of role.
- **Insist on planning and prework from all participants**. Especially when objectives are ambitious and the sense of urgency is great, participants will need to invest time preparing for a productive conversation. What can be sent ahead of time for review and comment? How can you poll the group for input in advance? What homework can you assign to sharpen the focus? Think about how you can create a level playing field in advance, enabling participants to launch right into a great conversation. Consider enlisting the aid of a senior executive if you suspect prework might be a tough sell.

- **Be specific about the type and amount of prep work required when you send out your meeting details,** and be realistic about how much time people are really likely to spend. For a two-hour call, you might realistically expect about an hour of prework. For a longer meeting, you can probably ask people to invest more time in preparation.
- **If you want people to do prep work, post or send content at least a few working days ahead of time.** Ask people to take a specific action as they prepare; for example, "Review the proposed budget and identify three places your team can cut expenses in the next three months" or "After reviewing competitors' data, be prepared to name the two areas we should be most worried about, with suggested ways to address them."

Table 10.4 (from Facilitate.com) provides several tips for making prework more compelling. Preparation and prework are needed, to

Table 10.4   Tips to Help Make Prework More Compelling

- **Provide incentives.** Offering a meaningful reward to those who complete their prework first (or on time) is an effective way to communicate that the prework has extrinsic as well as intrinsic value. Incentives can be fun and may help build a sense of team cohesion. Peer pressure to complete tasks on time can be more effective than reminders from the meeting facilitator.
- **Implement accountability.** Establish an explicit contract between you and your participants. Agree that you will design meaningful prework and deliver a session that meets stated objectives, and that they will come prepared to participate fully. How you do this depends on the type of session and group. You can't use the same techniques for an information-sharing meeting with a group of colleagues as you would for a senior management team.
- **Establish consequences for not completing the prework.** This can include disinviting participants after several unmet reminders or allowing them to "deselect" themselves from the session voluntarily. If this approach isn't appropriate for the group, you may need to ratchet down the objectives of the session, if people are not likely to come well prepared.
- **Whatever the consequences, be sure to communicate them early and clearly.** For groups that are meeting for one time, explain the prework requirement from the beginning and ask for acknowledgment as part of the registration process. For teams that meet frequently, initiate a conversation about prework agreements to help ensure meetings are productive. Revisit how well this is working and let the team manage the consequences for themselves.
- **Develop and execute a communication plan.** Preparing a communication plan for the prework is worth the effort for two good reasons. First, it is the vehicle for communicating the value, urgency, incentives, and consequences of doing the prework. Second, time spent in advance, building personal connections, is enormously valuable in creating a trustworthy, supportive team environment that enables true sharing and honest interaction. A bit of advance phone and e-mail work will pay big dividends.

Source: © Facilitate.com. Reprinted with permission. All rights reserved.

at least some extent, by all participants, regardless of the roles they play. The majority of meeting participants may need to review a report and submit a few questions in advance, as an example. Subject matter experts or presenters may need to come to the virtual table with specific content, such as answers to questions submitted in advance, or a graphical depiction of a new service model. An executive sponsor might need to be prepared to deliver a rousing three-minute pep talk at the beginning of a kickoff meeting, or a sober commitment to follow through on actions at the wrapup.

10.6 Importance of Thorough Preparation for Virtual Meeting Leaders

The meeting leader, of course, needs the most extensive preparation of all. Inasmuch as virtual meetings must be kept deliberately brief, with competing priorities beckoning participants bent on multitasking, given the opportunity, the virtual meeting leader can leave nothing to chance. All aspects of planning and facilitating virtual meetings must be exceptionally well orchestrated.

Table 10.5 is an example of a premeeting checklist created by Facilitate.com for virtual meeting leaders. Create your own premeeting list with the details that make the most sense to you and your team. For example, you may want to include URLs and other pointers to let participants know where they can access content to review in advance. You also may want to specify who's taking notes, who's facilitating the conversation, and who's keeping time for any given meeting, to make sure there are no surprises.

You may also want to identify "exclusions" (i.e., what this meeting is not about), to minimize opportunities for a last-minute disconnect, which can quickly derail a virtual meeting before it has a chance to get off the ground. For example, let's say the primary goal of the meeting is to agree on a new stakeholder communications plan in advance of a major new IT rollout. You may want to specify, in advance, that certain conversations will be excluded, or out of scope, such as detailed project planning, discussions about personalities and politics, or budget requests.

It's also helpful to mention assumptions in the meeting document you send in advance, asking people to take note and let you know if they have different ideas. This way, you stand a better chance of starting your virtual meeting with everyone on the same page at the

Table 10.5   Planning a Virtual Meeting—Checklist

- What is the primary purpose of this meeting?
- What are our desired outcomes, tangible and intangible? Which are the most important?
- What decisions or action items do we need to walk away with?
- What can we expect to be completed in 60 minutes or 90 minutes of allotted time?
- What do we need to let go or do at another time?
- How would I plan my agenda if this were a face-to-face meeting?
- How might we organize our conversation or activities into 15-minute segments?
- What level of participant interaction is important to each activity?
- What can we do asynchronously? What needs to be realtime?
- How can we limit the amount of information sharing during the meeting to maximize the amount of participant engagement and interaction?
- How might breakout groups or side conversations support participant engagement?
- Who needs to be involved at each step?
- Who can provide information ahead of time?
- Who needs to be informed afterwards?
- How can I best prepare participants?
- What technology will best support each step of the process? What's available?
- What handouts and worksheets will facilitate participant engagement and understanding?
- What materials can we send out in advance?
- What obstacles can I anticipate? How can I address these?
- What role do I need the client or sponsor to play?
- How much time will it take me to coordinate this meeting or series of meetings?
- What additional help do I need?
- Will this virtual process meet our objectives? What do I need to renegotiate?
- How will I evaluate the effectiveness of this meeting?

Source: © Facilitate.com. Reprinted with permission. All rights reserved.

outset. Using the example of a stakeholder communications plan in support of a new IT rollout, assumptions you may identify include

- All employees throughout the organization will be affected by this rollout, some more than others.
- We will pilot the new application in Singapore, Malaysia, and the Philippines from July 1 through July 31, with full-scale rollout to take place September 1 through January 31.
- Stakeholder types will vary by region, function, and organization.
- Some stakeholders are more important, and will be given more time and attention, than others.
- We will focus our meeting on the top three stakeholders, as agreed to by meeting participants in our premeeting poll.

10.7 Creating a Realistic Virtual Meeting Agenda

Start by knowing the kind of conversation you need to have. For example, is this a weekly status report meeting, where people will each provide a brief update and surface any outstanding issues? In this case, your agenda might be straightforward and easy to plot out. Or perhaps team members need to hash out some tough decisions that may be weighed down by heavy emotional content, which will require considerably more time, and perhaps multiple conversations. Think through what kind of conversations are needed, how they're likely to go, who needs to participate, and how easy or difficult it might be for you to manage the conversation successfully within the allotted time.

Also consider how many people will be part of the conversation, how well they know and like each other, the extent to which they're likely to be in sync, as well as language and culture differences. Make sure to set aside at least a few minutes for welcoming people and reviewing objectives, ground rules, assumptions, and exclusions as needed. Likewise, make sure to leave at least five to ten minutes for wrap-up, summary, and next steps. See Table 10.6 for a virtual meeting template.

Table 10.6 Virtual Meeting Template

PURPOSE:

OUTCOMES: PARTICIPANTS:

DATE/TIME:

TELECONFERENCE # AND PASSCODE:

ONLINE ACCESS: HTTP://_____

TIME (MIN)	AGENDA ITEM		ACTIVE/INACTIVE
5	Getting Started	Preparation:	
15		Preparation:	
		Engagement:	
15		Preparation:	
		Engagement:	
15		Preparation:	
		Engagement:	
10	Wrapping up—Next Steps	Action Items:	

Source: © Facilitate.com. Reprinted with permission. All rights reserved.

10.8 Tips for Designing a Successful Hybrid Meeting

A hybrid meeting is one where some people join from a conference room or someone's office and the rest join remotely from wherever they work. Remote participants often feel alienated or ignored, especially when they are in the minority. If it's absolutely essential for people to participate together from one location while others remain remote, here are some tips to consider for creating a more satisfying meeting experience for everyone:

- **Be sure the virtual participants can be heard.** Use a good-quality speakerphone, making sure it is well positioned and away from distracting table noises for people who call in. Make sure speakers are centrally placed. Test audio capability at the start of the call so you can calibrate whether in-person participants need to shift location when they speak.

- **Pay attention to making the virtual participants feel "present."** Try using a visual reminder of phone participants. For example, make a large tent card with the virtual participants' names and, ideally, their photos. Place them around the table, or around the phone, and make sure they can be seen by all. You can also try tying a balloon to the phone as a reminder that there are participants on the other end of the phone. When going around the room to ask for opinions or ideas, start with those on the phone to make sure you don't inadvertently leave them behind.

- **Compensate for nonverbal cues.** Ask all FTF participants to describe any activities that cannot be guessed at without visual cues such as writing on a flipchart, drawing a picture, or underlining important words on a flipchart. Make sure that you describe out loud any observations that convey important nonverbal communication, such as "Everyone is nodding" or "People are rolling their eyes."

- **Make sure that productive virtual participation does not depend on people seeing what is happening in the room**, unless you plan to use high-quality videoconferencing or some other means of "sight." For example, if people in the room are engaged in creating some type of graphics, you might send a photo file to virtual participants when they join the call, so you can all be seeing the same thing at the same time.

- **Limit the time that phone participants must spend in any given hybrid meeting.** Invite them to participate only at critical junctures, such as to give input at the start of the call, review the output of the group at the end, or to provide important perspectives related to a particular topic. It's unfair to everyone if virtual phone participants are left hanging for hours with little to do.

Here are some ideas for creating a level playing field in hybrid meetings:

- **Ask people in the room to say their names before speaking.** People on the phone should do the same. Exceptions: People with distinctive voices or accents, or people with whom everyone on the call is familiar probably won't need to state their names.
- **Try taking a digital picture of people in the room and sending to all virtual participants**, so they have a clearer picture of what (and whom) they're missing. Ask virtual callers to do the same.
- **Better yet: Ask someone to create a quick slide** with an oval in the middle to represent a table with photos of people and their names scattered around the table. Suggest that people keep a printout of this slide in plain sight during all calls, to create a stronger sense of team.
- **Eliminate side conversations**, which are distracting and discourteous to those trying to follow the conversation. This applies both to people in the room and on the phone. If you must have a sidebar, mute the phone briefly. Don't make this a practice, however, or you risk having others feel that they're being left out.
- **If participants in the room take a break, make sure people on the phone know exactly how much time they have before the meeting resumes.** You may want to mute the conference room phone until all participants are back. If people are late coming back into the room, let your phone participants know how much extra time they can take.

- **Mute selectively.** If some participants insist they must join the meeting via phone, despite the fact they don't have an obvious role, consider permitting them to listen in only, keeping the phone mute. (Of course, all on the call will need to know who's on the line, even if they remain entirely in listening mode.)

10.9 Setting and Enforcing Virtual Meeting Ground Rules

Ground rules, when enforced, help to keep meetings focused, on time, and on track. Virtual meetings, although they may share some ground rules with their FTF counterparts, have some unique variations. Here are some sample ground rules that may apply equally well for both FTF and virtual meetings. Consider which ones will be most important to live by when planning and running your own virtual meetings. Remember: once agreed to, ground rules should remain fairly stable for any given group. At the same time, participants should be flexible when certain ground rules don't fit a particular situation.

- **100% participation:** Everyone will stay focused on the conversation at hand and will avoid multitasking. This means putting aside any potential distractions and minimizing opportunities for disruptions during the meetings (e.g., close down all other apps, clear the pile of papers off your desk, and take one more minute to send that last burning e-mail).
- **No mute:** Keep yourself off mute, unless you're in an airport, call center, or otherwise have ambient noise. Staying off mute has a way of encouraging conversation and dissuading multitasking.
- **Identification:** Say your name before speaking, unless everyone knows each other's names by voice.
- **No hold**: Avoid using the "hold" function of your phone if you need to break away. Some phone systems play music when put on hold, which will immediately derail the group's conversation.
- **Handling off-topic issues/questions:** We will record on an electronic flipchart or some other shared document any topics, questions, or issues that can't be addressed during this meeting. We will allocate time at the end of this meeting to review next steps.

- **Staying focused and on track:** All participants help keep discussions focused by calling out when they see a discussion as being a digression. (This can be done by a show of hands if your meeting tool supports this capability, a quick poll, or simply by stopping and asking people if this topic is best covered now or later.) Ask someone to be timekeeper and let you know when time is almost up for a particular topic.
- **Punctuality:** The facilitator starts promptly and ends promptly. Latecomers and early departures take responsibility for catching up on their own time.
- **Share the air:** Be aware how much you have spoken compared to others, and be prepared to let others voice their opinions and offer their ideas. This means not interrupting people before they're through, and keeping track of participation, both yours and others. Raise your virtual hand to speak if several people seem to want to jump in.
- **Confidentiality:** What's said in the virtual room stays in the room, unless otherwise agreed to.
- **Clarity:** Be explicit about the treatment of legal issues, such as use of proprietary information, antitrust legislation, and so on. When people can't see what others are recording or writing, it's especially important to call this out.

A few more important points about the use and enforcement of ground rules during virtual meetings:

- **Enforce ground rules to keep people focused and on track.** Once people start to tune out, the meeting can quickly degrade into a chorus of half-hearted monosyllabic responses. Much better to establish ground rules right up front, preferably at least a few days prior to the meeting, so people can commit to being fully present during the whole session. As a result, you're far more likely to drive to a successful conclusion within the allotted time.
- **Set meeting ground rules and gain agreement from everyone.** Depending how countercultural some of the ground rules might be (e.g., no multitasking for an IT project team that's under pressure to put out dozens of fires each hour), send ground rules in advance so people can plan accordingly. Once

you gain agreement, reinforce ground rules vigilantly. The minute it's obvious that some people are multitasking or monopolizing the conversation, everyone will assume they can do the same.

- **Make sure ground rules are appropriate for all cultures.** For example, a ground rule of "all are peers" is not likely to gain traction in a culture where hierarchy and seniority matter deeply.
- **State ground rules about participation plainly at least twice: in advance and right before the meeting.** In your meeting preparation document, try writing something like: "The decisions we need to reach are crucial, so it's essential that you focus your full attention during this hour." You might also get specific and ask that people refrain from handling other tasks during the meeting.
- **When you begin the call, reiterate the need for everyone's full attention.** Ask if anyone has a special situation that requires her to take time away from the meeting, and when that temporary departure needs to take place, so that you can plan the discussion accordingly. By gaining verbal commitments from attendees right up front, they're less likely to slink away.

10.10 Guidelines for Great Global Team Meetings

Thanks to advances in technology, project team members scattered around multiple time zones work together as a matter of routine. However, without a keen understanding of important cultural differences that are most likely to affect collaboration, many virtual global project teams struggle to achieve their goals, or sometimes simply fall apart. Although some of these tips may be true for most virtual or cross-cultural teams, they are especially important for global project teams that rely on virtual communication as their primary means to collaborate.

I co-authored the following tips with my colleague Rich Johnston, an IT architect for UTC's Climate Controls and Security Systems in Syracuse, NY, who has honed his skills in leading cross-cultural project teams over the last several years. We wrote these tips with North American–based companies in mind, focusing on same-time meetings. However, these tips can be applied to globally dispersed organizations based anywhere. (Also refer to Chapter 7, "Navigating across Cultures, Time Zones, and the Generational Divide," for more tips related to leading global teams.)

- **Start with an unambiguous realistic agenda.** State what you plan to achieve in clear simple language. Especially if new members are joining, indicate that the meeting will be held in English. Build in sufficient time to allow nonnative English speakers to translate into their local language and back into English, which can take up to 50% more time than a native English speaker may need. Make it clear what you expect from each participant in the form of prework and participation during the meeting. Let team members know if substitutions or additions are acceptable, which is often the case if a strong command of English is required.
- **Establish and enforce meeting norms.** At the start of the meeting, summarize which countries, languages, and time zones are represented. Ask people to clear their desktops of any additional work during the call to allow for full and active participation by all. Remind people to speak clearly and avoid making interruptions. If you're using a web meeting tool, review the functions you plan to use, such as raising hands or sharing desktops. Make sure all know how to mute the phone, and remind people to say names before speaking. Indicate under what conditions team members may use instant messaging (or tweets). Remind people of the need to stay focused on the objectives, and indicate how you plan to capture and address "parking lot" issues that you won't have time to discuss during this meeting. Another norm that helps all feel equally valued regardless of location is to rotate meeting times to give everyone a chance to wake up at 5 a.m. or stay up until midnight.
- **Keep the language simple.** Use the fewest number of words to get your point across, which may require extraordinary preparation. Enunciate each word clearly, taking pains to pronounce them in a neutral accent. (This can be especially difficult for those with strong regional accents, but very critical for nonnative English speakers who may become quickly lost

when hearing a dropped "r" or a flat "a.") Avoid idioms and metaphors, which can confound or offend others. Americans in particular tend to use sports metaphors that have little or no meaning elsewhere, such as home run, out of left field, or slam-dunk.

- **Set the pace.** Allocate time for checkpoints at key junctures in the conversation. Pause periodically to allow silence to let all participants absorb what's just been said. Some people—Americans in particular—often feel compelled to puncture silence with a comment. For that reason, you may need to set a ground rule to ensure that people maintain these planned moments of silence. If you're using a web meeting tool, you can invite some participants to make comments in writing during these periods of reflection.

- **Engage all participants equally.** Many people can converse more easily by speaking and others by writing. Whenever possible, offer participants a chance to communicate in the ways they feel most proficient and comfortable. In addition to the phone, make use of web meeting technology that allows people to submit questions or offer responses in writing. People in some cultures may be reluctant to discuss sensitive or contentious topics out loud, especially where hierarchy is important. In this case, you may want to use a web meeting tool that allows for anonymity. Some people, whether due to culture or personality, may be reticent to speak. Make sure to go around the virtual table and solicit input from each team member. Be thoughtful about how best to pose a question that makes it safe for each to respond. Examples would be: what do you see as the greatest advantage/disadvantage of this solution? If you could change one thing about our proposal, what would it be?

- **Choose the best combination of tools.** Some meetings, such as a routine weekly status review, might be fine with just phone, as long as everyone has access to the needed documents. A business requirements discussion, on the other hand, would be most productive if people had multiple ways to get their ideas across, such as by writing on an electronic flipchart or posting notes for all to see. Videoconferencing can be especially valuable for new virtual team members who want to get a feel for each other's culture and working environment. When different time zones are involved, allow for asynchronous participation of some sort, such as by posting comments or questions in a virtual conference area whenever it's most convenient. Make sure that meeting notes are posted during the call as a way to verify for accuracy and understanding. Whatever the tool, make sure that all have reasonable access.

- **Identify and address miscues.** If you suspect that someone has responded to a conversation point in a way that suggests she has misunderstood a key point, acknowledge her comment and then proceed to paraphrase the original point and invite her to make an additional comment. If you have trouble following someone's accent, let him know you are having difficulty hearing him (rather than complaining that you can't understand his accent), and ask him if he can repeat his point a bit more slowly. If you still can't comprehend the point he is trying to make, you might try following up with him privately offline.

- **Use analogies for shared understanding.** If you believe that the information you want to convey may be overly complex, consider using an analogy that all can understand regardless of culture or native language. For example, when describing the actions to be completed prior to closing out a particularly complex project, you might use an analogy of a cargo ship leaving port, with all of the many tasks that have to be orchestrated in a certain sequence before the ship can push off. People can often connect best with a shared image, making it easier for them to agree on tasks, milestones, and dependencies.

Leading a global project team requires diplomacy, preparedness, superb listening skills, and the willingness to invest time in learning how cultural differences are likely to affect successful collaboration. Check in with a representative sampling of team members from time to time to hear how they're feeling and learn what improvements you and the team can make. Developing the needed skills and cultural literacy doesn't happen easily or fast, but once cultivated, can last a lifetime.

10.11 Summary

Whatever the goal—whether it's to make well-informed decisions, generate innovative solutions to vexing problems, or to create a shared strategy—all virtual meetings accomplish the most in the least time when the right people come together for a focused conversation.

Setting the stage for such conversations is particularly challenging when the participants all work from a distance, given that the margin of error is so small, and it's so easy to lose control once people tune out. For each key objective, consider what kind of conversation is called for, how long it might take, what preparation is needed, and who needs to be involved, and then plan accordingly.

11

KEEPING REMOTE
PARTICIPANTS ENGAGED

Even with the best-laid plans, not all of your virtual meetings will go as planned. Although some problems crop up due to a technological glitch, most often the toughest problems relate to the degree of engagement (or lack thereof) of meeting participants. The virtual meeting leader has to learn how to identify and address a number of tough challenges, all within a matter of seconds. Common challenges include:

- Some or all participants are very quiet, leaving the meeting leader to wonder what's causing the silence. Are they paying attention at all? (And if not, why not?) Are they upset? Bored? Angry? Doing e-mail?
- Some people dominate, whereas others give up trying to speak.
- Serious digressions threaten to take the group irrevocably off track.
- Lukewarm responses signal boredom or lack of interest (or perhaps multitasking).

In this chapter, we look at tips and techniques for maintaining balanced participation, keeping people engaged from afar, and dealing with dysfunctional behavior.

11.1 Understanding How and Why People Become Disengaged

The first step concerns the fact that virtual team leaders need to be able to detect why people have become disengaged before knowing what intervention or technique is likely to work best. Think about times when you have drifted in and out of virtual meetings yourself. We all try to pay attention during virtual meetings. We really do!

But then something diverts our attention (it doesn't take much) and we find ourselves tuning out, despite our best intentions.

It could be that a teammate just IM'd us with an urgent SOS for help. Or that our workload is too crushing to ignore as we put ourselves on mute and type away. Alternatively, it could be that the meeting is frankly just so dull that we stop paying attention, or that we really have nothing very valuable to add. Or maybe we were told we have to be here, when in fact there's nothing relevant for our work in any of the discussions.

It's a different story when we're the one trying to run a virtual meeting, however. When our participants become disengaged, we tend to become impatient and frustrated when others are not paying attention. After all, it's *our* meeting, and we think the conversation warrants attention. When you sense people have become disengaged, it's important to take a first best guess as to why some people have become disengaged in the first place, before you make an intervention.

You can pay a high price for getting it wrong. For example, if you gently suggest that people stop multitasking when in fact they're simply bored or offended, you may quickly lose their attention for the rest of the call at least, as well as credibility and trust. Here are some of the reasons virtual meeting participants typically become disengaged, followed by quick tips to help reel them back in.

- **People imagine they can do several things reasonably well, at the same time** (even though study after study has proven that no one can really bifurcate their frontal lobes very successfully). Establish ground rules up front, both in the meeting details you post or send in advance as well as at the start of the call. Remind people how important their participation is and ask them to clear their minds, desktops, and computer screens. Ask them to close down any applications they don't need for this call. Give them a minute or two to send that last e-mail, if need be. In exchange, you promise to keep everyone focused and on track.

- **People don't see the meeting as a good use of their time.** They're not sure why they're on the call, and have no idea how or when they'll be contributing. (And they have lots of other work they could be paying attention to right now.) To begin with, make sure you have invited the right people. The temptation is to invite everyone, because selecting the right people takes a lot of time and thought. Figure out who really has to be involved in this conversation, and find ways to involve others before or after the meeting. It's tough having a productive conversation with more than a handful of people, especially when a weighty decision must be made. You may need to come right out and ask people who seem particularly quiet if they feel this meeting is a good use of their time, without a hint of sarcasm. They may be very happy to make a graceful exit.

- **The conversation is drifting here and there, with little regard (or memory) for the stated objectives.** Make sure that you (and everyone else) have an agenda in mind and readily visible, so you can all track progress. Periodically refer back to the objectives, which should likewise be visible at all times. When you design the meeting, be realistic about how long each part of the agenda should take. If you find that you've miscalculated partway through a meeting, give people a choice to keep going on this topic, schedule a follow-up meeting, or park this topic for now. Appointing a timekeeper lets you focus on the group dynamics.

- **People are simply bored because there are limited opportunities to interact.** A good rule of thumb for virtual meetings is to allow for interaction about every five to seven minutes. Use virtual meeting technology to the fullest advantage. Build in opportunities for participants to multitask on task. Enable people to type in comments and then discuss. Encourage people to "talk across the table" with each other. Go around the virtual room to ask everyone to respond to a quick question relevant to the topic. Make it easy to pull everyone back in. If the group simply does not perk up quickly, be prepared to flex the agenda in the moment, or end the meeting early and build in more interaction the next time.

- **People are afraid to speak up.** This may be due to cultural differences (national, organizational, or functional), personal styles, demographics, the sensitivity of the topic, or any number of factors. Find ways to make it safe for everyone to speak up. Consider using web meeting tools that allow people to participate anonymously, either prior to or during the call. Structure questions in a way that elicits feedback without requiring criticism. For example, instead of asking participants what challenges they are struggling with, you might ask what one thing they'd like to see changed. You may need to reach out to certain people another way, either before or after your meeting, to hear their perspectives if you feel they have been reticent during the call.

- **The meeting does not inspire energy.** People may not be saying much because the meeting leader doesn't project enough energy and enthusiasm to get them talking. If you're leading a virtual meeting, make sure you project sufficient "pep" to act as group cheerleader. Stay focused on what's being said (or not being said), vary your tone and volume, and change the way you ask questions or otherwise involve the others in conversation.

- **Take a quick temperature check of the group** (this can be as easy as raising virtual hands) or ask people for a verbal quantification of their current energy level, on a scale of 1–10. If energy is low, ask everyone to stand up and move around to take a 30-second stretch break. Be prepared to shuffle the agenda around if that's what it takes to raise the energy level. (If your own energy level is low, drink some water, stand up, or otherwise shift the dynamics. When the facilitator's energy is low, you can "infect" others quickly.)

It's harder than ever to capture and maintain people's attention in any kind of in-depth conversation these days, even when we're speaking face to face. But when people can't see us and we can't see them, it's much more difficult to discern why people have become disengaged, making it infinitely harder to find just the right way to pull them back in. A well-thought-out meeting design, coupled with the intelligent use of tech tools and facilitation techniques, can work wonders to maintain focus and attention.

11.2 Discourage Multitasking with Clear Ground Rules and Focused Meetings

Try as you might, it's really tricky to enforce a no-multitasking ground rule. Much better to take responsibility for running a meeting that draws people in from the beginning and keeps them engaged throughout. Here are some tips:

- **Allocate meeting time appropriately.** Think about having people read documents in advance or listen to a virtual presentation so that you use the meeting time for discussion, debate, and decision making. Don't force people to listen to a boring presentation or review documents that could have been read more quickly in advance, leaving time for interactive conversation.
- **Stick to your stated agenda and strive to ensure that the team meets objectives through a well-directed conversation that's perceived as valuable by all.** If you allow a conversation to stray too far off course, you'll have trouble maintaining the kind of credibility you need to insist that people pay full attention. When you fail to fulfill a commitment as the meeting leader, you make it easier for others to give themselves a virtual "hall pass."
- **Keep the pace quick and discussions concise.** Nothing can drive someone to another task faster than a boring meandering conversation. You can ask for help right up front by indicating that you have a packed agenda, and that the only way you can help keep it focused is by having everyone else act as "co-facilitators" who can all help keep topics from straying too far off course.
- **Come prepared to keep people on their toes.** For example, have a list of questions to which you'll be seeking responses. (Some examples are: What are the top three challenges our sales teams are facing today? What is our single greatest opportunity to outpace our nearest competitor? What priority would you assign items in the following list? What do you like best/worst about X?) If you can let them know what you'll be asking in advance, all the better, because they're more likely to attend with thoughtful responses they're eager to share.

- **Vary the order and way in which you ask questions.** Keep people guessing to add a little boost of energy. For example, you might note the order in which people join the call, and assign each a number on a clock, starting with 1:00. For one set of questions, start with 1:00 and go clockwise (or counter-clockwise) until you hear a response from everyone. Another time you can ask for volunteers to go first, and check to see whether others have additional ideas not already covered. If you have a team photo, you can use this to go around the vir-tual room, noting who's participating and who's not. (Creating a team photo is as easy as asking people to post or send jpg files to you, which you can pop into a Word or PowerPoint (PPT) template, along with their names, seated around an imaginary table. I like to post these in our virtual conference area, and often also push them out in advance, with a request to print and have handy at the meeting, so everyone can be aware of the relative level of participation during a meeting.)

- **Try reversing the usual "all on mute except speaking" ground rule.** Ask everyone to stay off mute so they can be ready to participate instantly. This way, you'll hear any errant key clicks that reveal multitasking is occurring, and you can be prepared to make some well-timed comments. (This ground rule is not as viable under some circumstances, such as when participants work in very noisy environments, when there's a mysterious buzz that's impossible to locate easily, or when there are 15 or more people on the line.)

- **Use technology when possible to focus attention.** For exam-ple, a virtual classroom environment allows everyone to par-ticipate both via phone and within a web-based "classroom" area complete with whiteboard, a chatlike capability to pose questions, a chance to vote, ability to share applications, and even—if appropriate—throw tomatoes and award gold stars. The more dynamic and interactive the application, the less likely people are to toggle to an unrelated activity.

- **Pick up the pace.** If you sense that energy is dissipating or inter-est is fading, stop a minute to decide how best to proceed. If you're on a conference call, it's reasonable to say something like, "I am not hearing from many of you. I don't know whether this means we've lost you, or you're thinking what to say, or you're busy doing e-mail. Can someone help me understand what's going on?" Chances are, you will re-engage those you have lost. If you can, maybe it's time to speed up the conversation, determining (with participant agreement) whether it's OK to skip over certain topics in favor of devoting more time to others.

- **If participation is lagging, you might offer choices** such as taking a quick stretch break, checking in to see how people feel about the process, or reflecting back on the objectives and asking people whether they feel they're on track. Inject energy and project enthusiasm especially when group energy is low.

- **Know when to end.** Just because you've scheduled a two-hour meeting doesn't mean you have to use the whole time. If you've achieved most of your goals and people seem ready to end, let people know you're prepared to end the meeting early, and spend the last few minutes wrapping up loose ends, such as action planning or mapping out next steps. Review the objectives one final time and summarize what the group has done to achieve them. If there are outstanding questions or issues not yet addressed, gain agreement as to how best to move them forward. If, on the other hand, you realize you can't possibly achieve your objectives in this meeting, call it right away and discuss with the group how best to proceed. One option is to end the meeting now and reconvene another time, perhaps with different participants or with new information.

Despite your best efforts to keep people focused on the conversation at hand, one or two people may still find a way to multitask. If someone is diminishing the value of the conversation by habitual multitasking, you may have to come right out and say so. You'll need to consider how well you know the person, how formal or informal the meeting structure, the topic at hand, and your authority to control behavior, among other things.

You might try something such as (with humor): "Now Carol, can those telltale key clicks really be coming from you?" Or (more seriously) "Bill, I realize you have a ton on your plate now, but if you can just give us 15 more minutes of your undivided attention, we can let you go. Would that be OK?" The more you ignore the offending behavior, the more others are likely to assume that your tacit permission makes it OK for everyone to follow suit.

If worse comes to worst and you suspect that most meeting participants have moved on to something else, you simply may have to announce that you are ending the meeting, but before you do, you need to know when people can be prepared to spend focused time on completing the unfinished tasks. Try asking for a few smaller chunks of time, rather than trying for a longer period that may be difficult to schedule.

With everyone paying full attention, most phone-based meetings can end far more quickly, with better results. Your job is to persuade participants that their contributions are truly valuable, and that you plan to take your job of running a productive meeting very seriously. The more you can show that you're capable of leading the team to meet their objectives, the more likely they'll be to find other meetings in which to multitask.

11.3 Rx for Problem Participants Who Threaten to Derail Your Virtual Meetings

We've all suffered when the bad behavior of just one person can derail a whole meeting. Maybe it's been the know-it-all who steamrolls over anyone who tries to speak. Or the person who folds her arms and rolls her eyes without a word. It could be the guy who has nothing good to say about anything, or the one who keeps running in and out of the room to put out fires.

Even the most accomplished facilitator can be thrown off guard by problem participants. But sometimes all it takes is a perfectly worded phrase or a simple gesture that can help snap the meeting back on track. But when the meeting is virtual, you can't use those same nonverbal cues that can help shift behavior when you're sitting face to face.

Here are some tips and techniques for handling five of the most disruptive types of participants I have come across as a long-time facilitator of virtual meetings:

- **The Apparition:** He grunts occasional monosyllabic responses so you know he's on the call, sort of, but you get the sense he is not really there. You're not sure if he's muted the conversation so he can tend to his e-mail or if he simply has no interest. Whatever the reason for his divided attention, you need to pull him back in before everyone else follows suit. You can try saying his name to first get his attention before asking him a pertinent question: "Dave, in what ways were your experiences similar to Linda's?" Or you can announce that you'll be going around the virtual room to ask each person for a quick response to a burning question, even if you have to make one up on the fly. If he continues to sound distracted, state your observations straight up and ask for his help: "Dave, your input is crucial to this decision. Can we ask you to take part in the conversation for the next 15 minutes, so we make sure we've evaluated everyone's input?" Of course, your relationship with Dave and the rest of the team will determine how, exactly, you will call him out and ask for his help.

- **The Great Debater:** When everyone else seems about to agree on a key decision after a long discussion, the Debater goes on to regurgitate the very same options the group has assessed for the last hour. You suspect that she may have a hard time reaching closure, or that she may disagree with the decision (or she may simply enjoy being a contrarian). State your observations: "Jennifer, it sounds like you want to revisit an option that we've already discussed. I am not sure that further discussion will affect our vote. But let's ask the others." Poll the group to test your assumption. If some feel that Jennifer's option merits further discussion, set a firm time limit. If not, suggest that the group move to a vote. If Jennifer continues to object, implicitly or explicitly, you may need to ask her directly whether she feels ready to have the group make a decision. If Jennifer's commitment to the decision is critical to successful implementation, you may need to end the meeting without a vote and reach out to Jennifer 1:1 before you resume the discussion as a group. Otherwise, it's probably time to end the debate and simply vote.

- **The Curveball Pitcher:** Out of nowhere, she throws you a curveball, bringing up a topic that seems to have nothing to do with the conversation. You can't see whether everyone else is just as confused, and you suspect that if you let her go off on this tangent, the conversation will go way off track. When she pauses to take a breath, try jumping in to paraphrase her concern and bring the conversation back on track: "Kathy, it sounds like you want to make sure that our partners can link to our IT portal. Is that right? Great. Let's capture that in our notes for our next meeting. Now, we have 10 minutes left to finish brainstorming new ideas for our new campaign logo. Who has some more ideas?" This way, Kathy knows her concern has been heard and you have a plan to address it, and you've quickly invited others to rejoin the previous conversation. If she persists, you may need to enlist the group to help: "We have 10 minutes left to achieve our objectives, so we have a few choices. We can add 15 minutes to this call, schedule an additional meeting, or continue Kathy's topic at our next meeting. What would everyone like to do?" The majority will likely vote to park Kathy's topic in favor of a timely closure to the meeting.

- **The Steamroller:** Every time someone else tries to offer an opinion, this guy interrupts with his own ideas, many of which are repeated over and over. If you allow him to continue to hijack the conversation, you've lost the others for the duration of this call, and possibly long after. Because you can't give him the evil eye or kick him under the table, you need to be quick about yanking the virtual microphone out of his hands. Try succinctly summarizing his key points and acknowledging what you've all learned from his opinions. Then go around the virtual table and ask each participant to build on Jorge's ideas, or offer new ones based on their own experience. If Jorge still can't resist jumping in, firmly restate the meeting objectives and remind everyone about the ground rules all agreed to at the start of the call. If this isn't enough to rein Jorge in, try asking if he can take some time after the call to summarize his ideas via e-mail (or blog, wiki, etc.) and make his notes available for the group later on. If Jorge is a repeat offender, he may also need some 1:1 coaching after the call so he understands how his behavior affects the rest of the team.

- **The Buzz Kill:** Although everyone else on the call is revved up brainstorming new ideas, Laura responds critically to every idea before it's even fully formed, and offers no ideas of her own. Other participants suddenly shut down. To get the fountain of ideas flowing again, start by restating ground rules for brainstorming (e.g., quantity vs. quality, no critical comments, use of "and" vs. "but," etc.). Then summarize what you've just observed. "We had some great ideas flowing there for a while, and I noticed people suddenly stopped when their ideas were criticized. Can I ask everyone to withhold comments until we're through brainstorming? Let's restart with a new question." Alternately, you may need to direct your entreaty to Laura to make sure she understands how her behavior is affecting the team. If this is not the first time you've noticed how Laura's negative comments deflate the rest of the team, you might contact her privately after the meeting to help her reframe her comments in a way that encourages rather than discourages participation from others.

11.4 Summary

It takes sharp powers of observation and quick thinking to maintain continuous balanced participation throughout an entire virtual meeting. If you design your meeting for maximum engagement at the outset, the potential for people to disengage or go off track is far less. Even so, be prepared with an arsenal of interventions in case they're needed, including a set of stimulating questions that can be answered quickly by everyone.

12
TROUBLESHOOTING VIRTUAL MEETINGS

Normally, I ascribe to the 80/20 rule when it comes to planning virtual meetings. I know I can't predict every single problem that might rear its ugly head, so I do my best to anticipate and address those "gotchas" that are likely to happen about 80% of the time for any given situation. As far as the other 20% of the time, well, we'll just have to cross that bridge when we come to it. In the world of virtual meetings, we don't have very much time to cross that bridge before the entire meeting can go irredeemably off track. (The fact that we have no visual cues when things go awry makes it even harder to regroup and redirect the conversation.)

Here are some practical tips for anticipating and addressing problems that arise all too frequently during virtual meetings.

12.1 We Are Experiencing Technical Difficulties

Whether it's a bad phone line, a passcode that doesn't work, or a spotty network that boots us off at critical times, there are times when the technology does not work as planned. As meeting leaders, it's up to us to have a backup plan for whatever we expect can possibly go wrong. Here are some ideas:

- **Be ready to call an operator who can detect the source of that annoying buzz, or the participant who put the call on hold with 1980s music blaring.** This means making sure you know how to get an operator on the conferencing system you're using. Know how to use the keypad as directed by your phone service provider for any capabilities you may need, such as putting the phone on and off mute, leaving or entering the conference, turning on/off entry and exit chimes, and so on.

- **Give people your cellphone number (or your colleague's) if they need help dialing in.** (After all, your primary business line will be tied up.) Best to ask people to text you, so you won't have to interrupt your call out loud. If you and meeting participants all use the same IM system or chat function, use this channel as needed for "backdoor" communications that won't disrupt the rest of the call. (If you go this route, make sure to mute the audio on your device, so everyone can't hear the telltale "pinging" of messages to and fro.) Or, if your web meeting tool includes some kind of chat or notes function, you can instruct people to request assistance using these tools as well. Just make sure to let them know how to get help before everyone dials into the call, especially if the phone system may be one of the issues.

- **Test all passcodes before the meeting, just in case.** This applies to both phone conferencing as well as web meeting tools. Just because the passcode works one day, it won't necessarily work as planned the next day, for any number of reasons. For example, I experienced situations where the phone conferencing service had either reset or reallocated my passcode without notifying me in advance. (You can imagine the flurry of e-mails and IMs that transpired right before the call, to set people straight.) Of course, human error sometimes gets the best of us, where we either give out the incorrect passcode, or we have assigned a passcode confusingly similar to another conference. The net effect is that people arrive late and, usually, quite frustrated. Even when it's not the meeting leader's "fault," the meeting leader is ultimately accountable for such glitches, whether avoidable or not.

- **Always log on and dial in at least 10 minutes early,** to make sure there are no technical issues cropping up at the last minute. In addition to testing your dial-in and log-on information, check headsets, volume controls, and Internet connections. If you're using Skype or another means of IM, make sure it's activated.

- **Enlist a buddy to help run interference when needed.** This might be a co-facilitator, meeting participant, or another kind of assistant who can help solve technical issues while you lead the call. Decide in advance how you will communicate, whether it's via phone, IM, text, web meeting, or e-mail. Make sure you have his or her contact information at your fingertips. It never hurts to send a reminder about your meeting times, to make sure your buddy is accessible if you need a lifeline.

Above all, maintain a calm demeanor and a good sense of humor to help deflect even the most challenging situations. People may quickly forget about the snafu, but they are likely to remember your poise and grace under pressure.

12.2 Reconciling Time Zone Differences

A colleague and I recently led a webinar with people joining from several time zones. Initial meeting requests were sent by our sponsor, which we followed up with more e-mails, spelling out the start and end times for multiple time zones. We also posted the meeting time in different time zones very visibly in our virtual conference area. Nonetheless, three participants missed the call, assuming that the time was noon their local time (which happened to be Pacific Standard Time) as opposed to Eastern Standard Time. Lessons learned:

- **Use an official meeting request function, such as those available in Outlook or other calendar management systems, in addition to e-mail.** Most calendar management systems automatically convert start and end times into the local times for each invitee. Make sure to confirm that this conversion takes place accurately. (For example, I have one colleague for whom the meeting times I send out always show up as three hours earlier than they really are. We have no idea why this happens, but we do know to confirm the actual times with each other via e-mail.)
- **Clarify time zones.** If you are unable to send (or receive) a meeting request and have to rely instead on e-mail for meeting invitations, make sure to spell out the start and end times in your e-mail, indicating to which time zone you are referring. The rule of thumb is to start with the local time zone of most invitees. (Case in point: One of my colleagues can't seem to receive meeting requests when I send them from

Microsoft Outlook. Instead, she sees a screen full of nonsen-
sical gibberish. She now knows enough to ask me if I sent a
meeting request. And I, in turn, now realize that I will always
have to send invitations via e-mail in the future.)

- **When participants from multiple time zones are joining,
  state the meeting time in GMT (Greenwich Mean Time),
  as a minimum.** Look up the proper abbreviations for other
  time zones, and be prepared to spell some out. For example,
  CST is likely to be interpreted as Central Standard Time in
  the United States, and China Standard Time in Asia. Beware
  of summertime changes as well. Many countries have day-
  light or summertime hours, much as we have in the United
  States, which can wreak havoc on scheduling if the meeting
  planner isn't aware of the differences.
- **Provide a URL for your favorite time zone calculator** (e.g.,
  www.timeanddate.com), just in case there's any opportunity
  for confusion.

Despite your attention to detail, several people may still get the
time wrong. If those you are expecting are late joining the call, dis-
creetly send them a message (or have someone else do it) if you suspect
they have the time wrong.

12.3 Dealing with Uninvited Guests

Some people seem to feel more casual about inviting others to come
along to a virtual meeting rather than a face-to-face (FTF) meeting.
(After all, if hangers-on are very quiet, no one will know they are
there. Plus, the host doesn't have to order food or provide a chair, so
what's the harm?) Depending on the guests, this additional person
(or two, or three) may throw things off track in a hurry. Here are some
tips for handling this delicate situation:

- **Let your invited participants know in advance that they
  have been chosen with great care.** Indicate clearly whether
  replacements or delegates are permitted, and invite people to
  contact you if they feel that an important participant was left
  off the list, giving you time to assess the request. This way, if

the request makes sense, you can prepare the additional participant prior to your meeting.

- **If you're surprised with unexpected guests during the call, welcome them and indicate (diplomatically, of course) that you had not known they were coming.** This is not meant to chase them off the call exactly, but it's to let everyone know that you had prepared for this meeting with the invited participants in mind, and hints at the fact you're not able to alter the agenda easily to accommodate new people.
- **Let your surprise guests know they are welcome to stay** (only if that's true; otherwise you'll have to let them know how they can be caught up and by whom, after the call), and advise them that all participants invited in advance have been asked to do some preparation and planning to make the best use of the meeting time. This way, they're less likely to ask you to review what they missed in the prework.
- **If you don't know who invited these additional guests, you may want to ask** (e.g., "Hi Sarah, and welcome. I apologize that I had not realized you would be attending. Can you let me know who passed the invitation on to you, so I can make sure we're all in sync for future meetings?"). There may be a very good reason why Sarah has been asked to join this call, but unless you understand what the person who invited her had in mind, you have no idea what her role is intended to be, nor do you have any context for her contributions. She might be a vital contributor to the discussion, or she might be just as confused as you are as to why she showed up.
- **If you feel the extra participants' presence will not adversely affect the outcome of your meeting, you might just go ahead as planned** with an extra person or two, especially if their relationship is likely to be important over time.
- **If the additional participants might affect the outcome negatively, you may need to be clear about their level of participation,** which can range from none at all, to listening only, to partial participation at certain times. Be prepared to explain that you have carefully planned this meeting with a certain number and mix of participants in mind, and that to achieve meeting goals, you need to adhere to your design.

- **If you absolutely feel that the outcome of the meeting will be jeopardized by allowing the extra people to stay on, apologize that you can't extend an invitation just now,** and let them know that they will have other chances to participate, either during a future meeting or some other way (only if it's true, of course). Offer to follow up via phone as soon as possible after this call, so you can fill them in and mend any fences in need of repair.

Some people may be thrilled to be able to make a graceful exit from a meeting, giving them a welcome gift of time back in their schedule. Others may feel slighted that they were not enthusiastically welcomed. (After all, some may not even know they were not invited to begin with.) When in doubt, prepare to reach out to your surprise guests afterwards, making amends if needed.

12.4 Handling People Who Show up Unprepared

Many people simply don't take the prework seriously ("Gosh, I've just been flat out!"). Some do this as a matter of course, and others gloss over the prework just once in a while. How you handle this transgression depends on a variety of factors, including your relationship with the participant, his track record for doing prework, and the importance of doing this particular prework to achieve your desired outcomes.

Whatever the reason, if someone shows up to your virtual meeting without having done the needed prework, here are some tips:

- **Resist the temptation to spend a few minutes of everyone else's time catching people up on what they missed.** (Of course, this depends to some extent on your relationship to those people.) The participants who did manage to invest their time doing the prework will be grateful that you have respected their time by not rehashing content for those who did not come fully prepared to contribute.
- **Remind everyone how the prework was intended to help people prepare for a productive conversation today.** (For example: "Everyone was asked to read the Q4 business plans for each region and come to the table with the top three recurring themes

to help us agree on priorities for the coming year. Because we only have one hour for today's meeting, we agreed that the pre-work was essential to meeting our goals in such a short time.")

- **Staying with this example, let everyone know you'll be starting the conversation with those who have come ready** to identify the recurring themes they noted as they read through the business plans. Those who did not do the prework may either listen to the others for now, or they can take a few minutes out to read the business plans, and then rejoin the discussion when they're ready. Either way, everyone now knows that there are consequences for not being prepared, making it less likely that you'll see repeat offenses.

- **Before you end your current meeting, be ready to describe the prework for next time, and explain how it will be used.** Also, give people an idea of about how long you expect it to take. This way, everyone knows what the prework consists of, why it's important for them to do it, and how much time they'll need to put aside to get it done. Those who did not come prepared this time won't want to be caught out again.

12.5 Keeping Remote Participants Feeling Connected

Let's say you're leading a meeting where many of the participants are physically together, whereas a few others are participating remotely, including you. As the meeting proceeds, you notice that the people huddled in the conference room are repeatedly putting themselves on mute. You assume this is so they can have a side conversation or cover up the fact that they're multitasking. Whatever the reason, you and other remote participants feel adrift, cut off from what could be an important conversation, and unable to engage the others no matter how hard you try. You have a lot to accomplish in a short time, and you won't have another chance any time soon to reconvene.

Here are some tips to try to keep remote participants feeling connected when they feel ignored or left out in the midst of a virtual meeting:

- **First, explain how you feel as a remote participant when you hear dead space for several minutes at a time.** (For example: "I'm not sure whether you guys are having a sidebar conversation, doing other work, or whether you've all gone off for a coffee break and forgotten to tell us. All I know is that I feel like I'm missing an important conversation that I'd like to be part of.")

- **Ask the other remote participants how they're feeling, assuming they have not already turned their attention to other work.** Ask the question in such a way that other participants don't feel overly embarrassed to respond candidly (e.g., "Bob, what were you imagining was going on for the last few minutes when we didn't hear from the folks in Boise?"). These responses should help heighten the awareness of conference room participants as to how their silence affects team members who cannot see what is actually going on.

- **Request that all participants stay off mute for the duration of the meeting, to make sure everyone can join the conversation equally.** (This ground rule applies to both in-person and offsite participants.) Suggest that "off mute" be a new ground rule going forward, with some exceptions for those surrounded by ambient noise or other audible distractions.

- **Continue the conversation by first engaging remote participants before moving to those in the conference room,** signaling to remote participants that you regard their thoughts and opinions as just as important as everyone else's, if not more so at this point, given that they did not have an equal shot at joining the conversation earlier.

- **If you hear significant audio distortions from those speaking from the conference room,** as often happens when a speakerphone is used, be prepared to call a five-minute break to allow conference room participants to find a place where they can dial in from a private location, unless such a move is downright impossible.

- **Before this call ends, ask all conference room participants to dial in remotely from a quiet location for future meetings,** explaining that you want to make sure you create as level a playing field as possible, given that some people work together and others apart. If there are good reasons that a particular group needs to meet face to face, suggest they meet as part of the whole team remotely, along with everyone else. They can then convene in a common area for other needed discussions at a different time. (This may not always be possible or practical, but it's worth suggesting ways that everyone works hard to create a level playing field in all team conversations, regardless of where they work.)

12.6 Summary

Despite the best-laid plans, unexpected situations will test the mettle of even the most seasoned meeting leader. With virtual meetings, you have less time to get things right. As you're planning your next meeting, make notes about every detail that can possibly trip you up and factor these into your design. By planning for the worst case, you'll get the best results.

13

SUMMARY

Now that you have these new tips and tools under your belt, are you ready to blaze new paths of glory as the best virtual team leader who ever lived? Do you feel you're completely prepared to run virtual meetings that draw people in, keep them thoroughly engaged, and get them coming back for more? No? Not quite yet?

Most people need a little (or a lot) of practice to bolster confidence and hone new skills, especially when they themselves don't have role models in their line of sight. (After all, many virtual team leaders work remotely from their own managers, depriving them of the opportunity to observe "best practices" leadership themselves, much as their own team members miss seeing you in action every day.)

You can start applying the tips and techniques you've learned in this book in different ways. One approach is to perform a gap analysis to take stock of which skills and knowledge you most need to become a successful virtual team leader. For example, do you need to ramp up your listening skills to lead more effectively in an environment where you have few visual cues to go by? Can you use help facilitating difficult conversations across time and distance? Would you benefit by knowing how to more deftly navigate through cross-cultural differences? Pick a few areas where the gap between your current and desired future state is greatest, and map out a development plan for yourself. Enlist the help of colleagues, mentors, managers, or HR if needed.

For example, if you know you need to beef up your facilitation skills, you can take one of many paths. You can seek out courses, either within the organization, outside, or online. Look for books, blogs, websites, or articles that pertain to your unique situation. Identify people in your organization who you regard as exceptional facilitators and listen in (with permission from all, of course) to learn what makes them so effective. Then, ask them to shadow you and give you a candid assessment of where you can most learn to improve. Solicit feedback from your team members and colleagues periodically—1:1, as a team, via survey, or some combination—to gauge your facilitation effectiveness. For example, you might probe to determine whether people feel meeting time is well spent, whether all opinions are being heard, conversations are relevant, or whatever other metrics make the most sense for your team. Be reasonable about how many new skills you can cultivate at any given time, and make sure to allocate the time and resources you need, either during work or outside work, for learning and practice.

Another approach to applying what you've learned in this book is to start with one or two key challenges that keep you (or your team members) up at night. Maybe you have a burning need to establish credibility as team leader or to repair broken trust across the team. Or perhaps you have a challenging task you need to start now, such as designing an interactive virtual business requirements planning session or helping the team decide how they can make better use of a team portal.

Pick a place to start and decide what changes you or your team need to make, and what work it will take to get you there. For example, if you need to cultivate trust across the team, you might start by discovering what attitudes and behaviors members consider as especially trustworthy. This can vary by the individual, national culture, phase of the project, or a variety of other factors. For example, some might see empathy from other team members as especially important to building trust, whereas others might regard dependability or sincerity as more important. (Naturally, you'll need to find a way to ask team members in a way that elicits the right information, which can take the form of a 1:1 phone interview, a team activity, an online survey, or a combination.)

Based on what you learn, design activities that can help team members build trust as a natural part of their collaboration. Some examples are to ask each person to reveal his or her top challenge for the coming week to kick off each team meeting, or to end by sharing one new idea everyone can use. Pair members up by task to jumpstart relationship-building where most needed. Invite people to share special gifts and to acknowledge areas that need strengthening, and encourage people to cross-pollinate knowledge and skills.

If the team is suffocating from information overload, it might be time to help guide members in creating norms about document-sharing. The first step is to schedule a same-time virtual meeting to brainstorm the perceived pros and cons of using the existing portal and identify barriers to participation. For example, if some people say they find the process too cumbersome or confusing, start with a norm that pertains to ease of use or simplicity of navigation. Use templates included in this book to help jumpstart this process.

As you make changes and try new approaches, in some cases you may want to be transparent about the goals you're trying to achieve or the changes you're trying to make. In other cases, you may want to develop skills or make changes that you prefer not to announce, at least not just yet, such as striving to make all team members feel equally valued or cultivating credibility for yourself as team leader.

Becoming an exceptional virtual team does not happen overnight. Even the most seasoned leader who's completely comfortable in leading successful co-located teams can struggle, at least occasionally, when it comes to inspiring, motivating, and engaging virtual teams. Start anywhere on your journey to becoming an exceptional virtual leader. It can be a baby step or a giant leap. You might need some level of formal training and a reliable safety net as you step out of your comfort zone, or you might do best by making changes in small steps, using trial and error as your best teacher.

As you move ahead, tap into resources, such as websites from Guided Insights (www.guidedinsights.com) for articles, tips, and guides. (I publish a free monthly e-zine, *Communique,* which contains practical tips for virtual team leaders, team members, and others who want to thrive in the virtual global world.) Many other authors and consultants are incredibly generous when it comes to sharing articles, white papers, tips, guides, and online publications.

A great way to get just-in-time answers to your current conundrums is to join subgroups that focus on virtual leadership and collaboration within social marketing sites such as LinkedIn, Twitter, Facebook, and others. Search online resources from academia, magazines, blogs, consultants, and organizations that offer tips and skills. Looking within, seek out other virtual leaders or facilitators within your organization with whom you can set up a "buddy system" where you can learn, grow, and develop new virtual leadership skills with peers and colleagues.

There's nothing really magical about learning how to galvanize, mobilize, motivate, and engage a team of people across time and distance. Most of the leadership skills you've already honed can be transferred, to a greater or lesser degree, to the virtual world. With a few new skills, thoughtful planning, and the courage to try out new tips and approaches, you'll be well on your way. The results just may be nothing short of magical.

Index

A

Accountability, 26–27
    avoidance of, 140
    mutual, 93
    virtual professional development
        and, 81
Achievements
    cross-cultural differences in
        recognition of, 117–118
    recognition of, 33–34, 95–96,
        97–98
    virtual celebrations of, 98–101
Active listening, tips for team
    leaders, 85–88
Active participation
    operating principles supporting,
        47–48
    virtual team meetings, 157
Agendas
    creation of for global team
        meetings, 174
    creation of for virtual meetings,
        150, 154, 156–158, 169

Alignment, 2
Aptitudes, assessing, 13–14
Assignments, 30
Assimilation of cultural differences,
        112–116
Asynchronous communication,
        30, 60
    inclusion of in communications
        planning, 56
    use of for brainstorming, 71
    use of for introductions, 66–67
    virtual meetings, 147
Attitudes
    micromanagement and, 134
    virtual team leaders, 5

B

Behaviors
    modeling of, 131, 138
    virtual team leaders, 5
Brainstorming, 70–73
    creative thinking and, 90–91
    real-time conversations, 75